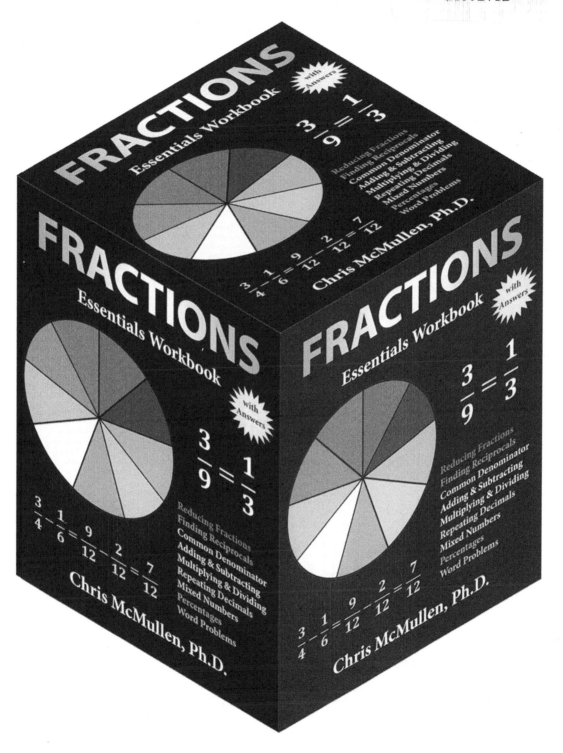

FRACTIONS

Essentials Workbook

with Answers

$$\frac{3}{9} = \frac{1}{3}$$

$$\frac{3}{4} - \frac{1}{6} = \frac{9}{12} - \frac{2}{12} = \frac{7}{12}$$

Reducing Fractions
Finding Reciprocals
Common Denominator
Adding & Subtracting
Multiplying & Dividing
Repeating Decimals
Mixed Numbers
Percentages
Word Problems

Chris McMullen, Ph.D.

Fractions Essentials Workbook with Answers
Improve Your Math Fluency
Chris McMullen, Ph.D.

www.improveyourmathfluency.com

Zishka Publishing

ISBN: 978-1-941691-25-0

Textbooks > Math > Fractions
Study Guides > Workbooks> Math
Education > Math > Fractions

CONTENTS

INTRODUCTION

This workbook is designed to help understand essential fractions concepts and practice essential fractions skills. Each chapter focuses on one main topic such as how to find a common denominator or how to convert a mixed number into an improper fraction.

Every chapter begins with a concise explanation of the main concept, followed by a few examples. The examples are fully solved step-by-step with explanations, and should serve as a valuable guide for solving the practice problems. The answer to every practice exercise is tabulated at the back of the book.

The workbook begins with a pretest assessment and concludes with a posttest assessment. The idea behind the pretest is to assess your understanding before you begin using the workbook. When you finish the workbook, comparing the posttest with your pretest assessment can help you see how much you have improved. Teachers may also find the pretest and posttest useful for evaluating student improvement.

A variety of essential fractions skills are covered in this workbook. The first chapter starts out simple by showing how fractions can be represented visually, and the difficulty of the lessons progresses as the book continues. Students will learn how to do arithmetic with fractions, how to convert to decimals or percentages, how to find repeating decimals, and how to solve word problems that involve fractions.

Chapters 16-18 challenge the student by expressing a variety of problems with mixed numbers (whereas the earlier chapters use improper fractions instead). The posttest exercises also challenge the student to apply a variety of skills learned in this book.

May you (or your students) find this workbook useful and become more fluent with these essential fractions skills.

PRETEST ASSESSMENT

The purpose of this pretest is to help you (or a teacher or parent) see how much you already know about fractions—if anything. It's okay if you score low on the pretest: It just means that you haven't learned much about fractions yet. The goal of this workbook is to help you learn. If you score low on the pretest, it shows that you have much to gain by using this workbook.

A secondary goal of this pretest is to show you a sample of the kinds of fractions skills that you will learn in this workbook. If you don't know how to solve a problem on the pretest, don't worry: You will learn how to solve it later in this book.

When you finish the pretest, check your answers with the key on page 10. After you complete this workbook, taking the posttest and comparing with your pretest score will help you see how much you have improved.

❶ Express $\frac{15}{40}$ as a reduced fraction.

❷ Which is bigger, $\frac{2}{3}$ or $\frac{5}{8}$?

❖ Continued on the next page.

❸ What is the reciprocal of $\frac{4}{7}$?

❹ What is the reciprocal of 5?

❺ Evaluate $\frac{3}{4} + \frac{1}{6}$. Express your answer as a reduced fraction.

❻ Evaluate $\frac{3}{5} - \frac{2}{9}$. Express your answer as a reduced fraction.

❼ Evaluate $\frac{2}{3} \times \frac{5}{4}$. Express your answer as a reduced fraction.

❖ Continued on the next page.

❽ Evaluate $\frac{5}{12} \div \frac{3}{4}$. Express your answer as a reduced fraction.

❾ Express 0.75 as a reduced fraction.

❿ Express $\frac{5}{8}$ in decimal form.

⓫ Express 0.4 as a percentage.

⓬ Express $\frac{5}{2}$ as a percentage.

❖ Continued on the next page.

⓭ Express 35% as a reduced fraction.

⓮ Express 150% in decimal form.

⓯ Express $\frac{8}{3}$ as a repeating decimal.

⓰ Express the repeating decimal $0.\overline{45}$ as a reduced fraction.

❖ Continued on the next page.

⓱ Express the improper fraction $\frac{20}{3}$ as a mixed number.

⓲ Express the mixed number $2\frac{1}{4}$ as an improper fraction.

⓳ Evaluate $3\frac{2}{5} + 5\frac{3}{4}$. Express your answer as a mixed number.

⓴ A parking lot has 6 red cars and 12 blue cars. What fraction of the cars are blue?

❖ Check your answers on the next page.

Check your answers to the pretest.

1 $\frac{3}{8}$ **2** $\frac{2}{3}$ **3** $\frac{7}{4}$ **4** $\frac{1}{5}$ **5** $\frac{11}{12}$

6 $\frac{17}{45}$ **7** $\frac{5}{6}$ **8** $\frac{5}{9}$ **9** $\frac{3}{4}$ **10** 0.625

11 40%   **12** 250%   **13** $\frac{7}{20}$ **14** 1.5   **15** $2.\overline{6}$

16 $\frac{5}{11}$ **17** $6\frac{2}{3}$ **18** $\frac{9}{4}$ **19** $9\frac{3}{20}$ **20** $\frac{2}{3}$

If you scored low on the pretest, don't worry: This just shows that you can learn much by using this workbook. These are skills that we will learn. They aren't things that you already need to know.

Are you wondering how we got one of the answers? We'll show you after you complete the posttest at the end of this book. You will learn these skills while using this workbook, and the posttest will help you see how much you have learned. We recommend that you don't read the explanations for the posttest until you've completed this workbook and have attempted the posttest.

However, page 185 does offer a few notes that may be helpful in the meantime. If you feel that you absolutely must check the full explanations to the pretest now, you can find them on page 245.

1 PIE SLICES

One way that a fraction can be helpful is if you have less than a whole object. For example, if you serve a few slices of a pie, what remains is a fraction of the pie.

It's convenient to draw slices of a pie to help represent fractions. As an example, the shaded pie slices below represent the fraction $\frac{5}{8}$ because 5 out of the 8 slices are shaded gray.

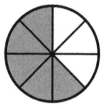

A fraction has a numerator and a denominator. The number at the top of the fraction is called the numerator, while the number at the bottom of the fraction is called the denominator.

$$\frac{numerator}{denominator}$$

For example, in the fraction $\frac{5}{8}$, the numerator is 5 and the denominator is 8.

Definitions

- **numerator**: the number that appears in the top of a fraction.
- **denominator**: the number that appears in the bottom of a fraction.

Example: What fraction of the pie shown below is shaded gray?

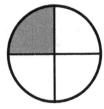

- The numerator is 1 because 1 slice is shaded gray.
- The denominator is 4 because there are 4 slices in total.

The fraction of the pie that is shaded gray is $\frac{1}{4}$.

Example: What fraction of the pie shown below is shaded gray?

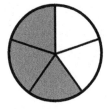

- The numerator is 3 because 3 slices are shaded gray.
- The denominator is 5 because there are 5 slices in total.

The fraction of the pie that is shaded gray is $\frac{3}{5}$.

Example: What fraction of the pie shown below is shaded gray?

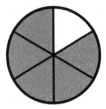

- The numerator is 5 because 5 slices are shaded gray.
- The denominator is 6 because there are 6 slices in total.

The fraction of the pie that is shaded gray is $\frac{5}{6}$.

Date: _____   Score: ___/9   Name: _____

Chapter 1 Exercises – Part A

Directions: For each pie shown below, what fraction of the pie is shaded gray?

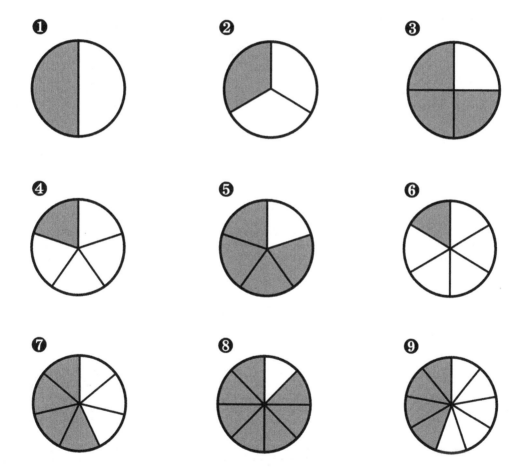

❖Check your answers using the Answer Key at the back of the book.

Date: _____   Score: ___/9   Name: _____

Chapter 1 Exercises – Part B

Directions: For each pie shown below, what fraction of the pie is shaded gray?

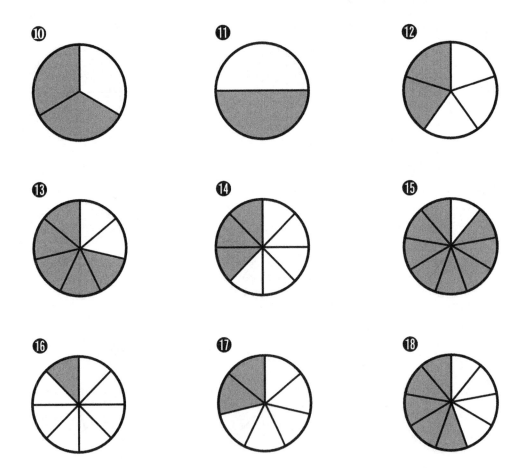

❖ Check your answers using the Answer Key at the back of the book.

2 REDUCING FRACTIONS

Some fractions can be reduced. For example, the fraction $\frac{4}{8}$ reduces to $\frac{1}{2}$. You can see this visually in the two pies below. In the left pie, 4 out of the 8 slices are shaded gray, while in the right pie, 1 out of the 2 slices are shaded gray. In each case, exactly one-half of the pie is shaded gray, showing that $\frac{4}{8}$ is equivalent to $\frac{1}{2}$.

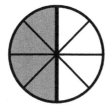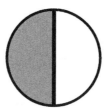

A fraction is reduced when it isn't possible to find an equivalent fraction with a smaller numerator and denominator (using whole numbers for both the numerator and denominator). For example, the fraction $\frac{3}{5}$ is a reduced fraction because you can't find a simpler fraction that is equivalent to $\frac{3}{5}$, whereas the fraction $\frac{3}{9}$ is reducible because the simpler fraction $\frac{1}{3}$ is equivalent to $\frac{3}{9}$. In the diagram below, you can see that whether you shade 1 out of 3 slices gray or shade 3 out of 9 slices gray, either way one-third of the pie is shaded gray.

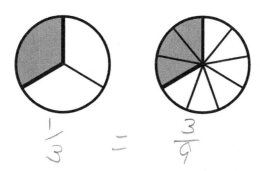

Definition

- **reduced**: a fraction that is already in the simplest possible form. You can't find an equivalent fraction with a smaller numerator and denominator.

It isn't necessary to draw pictures of pie slices to find the reduced form of a fraction. A fraction can be reduced by dividing both the numerator and denominator by a common factor.

For example, consider the fraction $\frac{12}{18}$. The numerator (12) and denominator (18) are both multiples of 6. If we divide the numerator and denominator by 6, we will obtain an equivalent fraction. As shown below, $\frac{12}{18}$ reduces to $\frac{2}{3}$.

$$\frac{12}{18} = \frac{12 \div 6}{18 \div 6} = \frac{2}{3}$$

In the previous example, we applied the following rule: If you divide both the numerator and denominator of a fraction by the same number, you get an equivalent fraction. We divided 12 and 18 both by 6 in the example above.

Note that 12 and 18 are both divisible by 6. Put another way, 6 is a factor of both 12 and 18.

Factors are numbers that can be multiplied together to make a larger number. For example, 2 and 5 are factors of 10 since $2 \times 5 = 10$.

Since 12 can be written as $12 = 2 \times 6$ and since 18 can be written as $18 = 3 \times 6$, this is why the number 6 is a factor of both 12 and 18. This is why we were able to divide both the numerator and denominator of $\frac{12}{18}$ by 6 to reduce $\frac{12}{18}$ down to $\frac{2}{3}$ in the previous example.

The following page provides a review for how to find the factors of a number.

Definitions

- **factors**: numbers that can be multiplied together to make a larger number. For example, 5 and 7 are factors of 35 since $5 \times 7 = 35$.
- **numerator**: the number that appears in the top of a fraction.
- **denominator**: the number that appears in the bottom of a fraction.

Finding Factors

Numbers that multiply together to make a larger number are called factors. For example, 3 and 5 are factors of 15 since $3 \times 5 = 15$.

One way to determine the factors of a number is to think of different ways to multiply smaller numbers together to make that number. For example, consider the number 21. The factors of 21 are 3 and 7 (since $3 \times 7 = 21$).

Following are some tips for how to find common factors.
- 2 is a factor if a number is even. (An even number ends with 0, 2, 4, 6, or 8.) For example, 2 is a factor of 14 since 14 is even. Note that $2 \times 7 = 14$.
- 3 is a factor if the digits of a number add up to a multiple of 3. For example, 3 is a factor of 51 because the digits of 51 add up to 6 (since $5 + 1 = 6$), and since 6 is a multiple of 3 (since $3 \times 2 = 6$). Note that $3 \times 17 = 51$.
- 4 is a factor if the last two digits of a number are a multiple of 4. For example, 4 is a factor of 140 since the last two digits (40) are a multiple of 4 (since $4 \times 10 = 40$). Note that $4 \times 35 = 140$.
- 5 is a factor if a number ends with 5 or 0. For example, 5 is a factor of 45 because 45 ends with 5. Note that $5 \times 9 = 45$.
- 6 is a factor if the digits of a number add up to a multiple of 3 and if the number is even. For example, 6 is a factor of 18 because the digits of 18 add up to 9 (since $1 + 8 = 9$), 9 is a multiple of 3 (since $3 \times 3 = 9$), and since 18 is even (it ends with 8). Note that $6 \times 3 = 18$.
- 8 is a factor if the last three digits of a number are a multiple of 8. For example, 8 is a factor of 1400 since the last three digits (400) are a multiple of 8 (since $8 \times 50 = 400$). Note that $8 \times 175 = 1400$.
- 9 is a factor if the digits of a number add up to a multiple of 9. For example, 9 is a factor of 549 because the digits of 549 add up to 18 (since $5 + 4 + 9 = 18$), and since 9 is a multiple of 18 (since $9 \times 2 = 18$). Note that $9 \times 61 = 549$.
- 10 is factor if a number ends with 0. For example, 10 is a factor of 50 because 50 ends with 0. Note that $10 \times 5 = 50$.

In the examples that follow, we will reduce fractions by finding factors that are common to both the numerator and denominator.

Example: Reduce the fraction $\frac{5}{10}$ down to its simplest form.

Note that 5 and 10 are both divisible by 5:
- $5 \div 5 = 1$
- $10 \div 5 = 2$

This means that 5 is a common factor of the numerator and denominator. Divide the numerator and denominator both by 5 in order to reduce the fraction.

$$\frac{5}{10} = \frac{5 \div 5}{10 \div 5} = \frac{1}{2}$$

The fraction $\frac{5}{10}$ reduces down to $\frac{1}{2}$.

Example: Reduce the fraction $\frac{4}{6}$ down to its simplest form.

Note that 4 and 6 are both divisible by 2:
- $4 \div 2 = 2$
- $6 \div 2 = 3$

This means that 2 is a common factor of the numerator and denominator. Divide the numerator and denominator both by 2 in order to reduce the fraction.

$$\frac{4}{6} = \frac{4 \div 2}{6 \div 2} = \frac{2}{3}$$

The fraction $\frac{4}{6}$ reduces down to $\frac{2}{3}$.

Example: Reduce the fraction $\frac{21}{35}$ down to its simplest form.

Note that 21 and 35 are both divisible by 7:
- $21 \div 7 = 3$
- $35 \div 7 = 5$

This means that 7 is a common factor of the numerator and denominator. Divide the numerator and denominator both by 7 in order to reduce the fraction.

$$\frac{21}{35} = \frac{21 \div 7}{35 \div 7} = \frac{3}{5}$$

The fraction $\frac{21}{35}$ reduces down to $\frac{3}{5}$.

Example: Reduce the fraction $\frac{24}{30}$ down to its simplest form.

First we will show a long solution, and then we will show a short solution. It will be instructive to compare the two methods.

Long solution: Note that 24 and 30 are both divisible by 2:

- $24 \div 2 = 12$
- $30 \div 2 = 15$

This means that 2 is a common factor of the numerator and denominator. Divide the numerator and denominator both by 2 in order to reduce the fraction.

$$\frac{24}{30} = \frac{24 \div 2}{30 \div 2} = \frac{12}{15}$$

Although the fraction $\frac{24}{30}$ reduces to $\frac{12}{15}$, we're not finished because $\frac{12}{15}$ can be reduced further. Note that 12 and 15 are both divisible by 3:

- $12 \div 3 = 4$
- $15 \div 3 = 5$

This means that 3 is a common factor of 12 and 15. Divide 12 and 15 both by 3 in order to reduce $\frac{12}{15}$.

$$\frac{12}{15} = \frac{12 \div 3}{15 \div 3} = \frac{4}{5}$$

The fractions $\frac{24}{30}$ and $\frac{12}{15}$ both reduce down to $\frac{4}{5}$. The final answer is $\frac{4}{5}$.

Short solution: Note that 24 and 30 are both divisible by 6:

- $24 \div 6 = 4$
- $30 \div 6 = 5$

This means that 6 is a common factor of 24 and 30. Divide 24 and 30 both by 6 in order to reduce the fraction completely in a single step (instead of two steps).

$$\frac{24}{30} = \frac{24 \div 6}{30 \div 6} = \frac{4}{5}$$

The fraction $\frac{24}{30}$ reduces down to $\frac{4}{5}$. We achieved this final answer in one step.

Compare: 6 is the greatest common factor of 24 and 30. Your solutions will be shorter if you can think of the largest factor that is common to the numerator and denominator. (If not, that's okay, but you may need to complete more steps.)

It's possible for a fraction to reduce to a whole number. For example, the fraction $\frac{6}{2}$ reduces to 3. That's because 6 divided by 2 equals 3. (You can think of the fraction line as being a division symbol. That is, $\frac{6}{2}$ is equivalent to $6 \div 2$.)

Here is a special case: If the denominator of a fraction equals 1, the fraction reduces to the numerator. For example, $\frac{2}{1}$ reduces to 2 (since $2 \div 1 = 2$).

Only apply this rule when the bottom number (the denominator) equals one. There isn't a rule like this for when the top number (the numerator) equals one. For example, the fraction $\frac{2}{1}$ reduces to 2, but the fraction $\frac{1}{2}$ doesn't reduce at all.

Example: Reduce the fraction $\frac{15}{3}$ down to its simplest form.

Note that 15 and 3 are both divisible by 3:
- $15 \div 3 = 5$
- $3 \div 3 = 1$

This means that 3 is a common factor of the numerator and denominator. Divide the numerator and denominator both by 3 in order to reduce the fraction.

$$\frac{15}{3} = \frac{15 \div 3}{3 \div 3} = \frac{5}{1} = 5$$

The fraction $\frac{15}{3}$ reduces down to 5. We applied the rule that $\frac{5}{1}$ is equivalent to 5.

Example: Reduce the fraction $\frac{14}{7}$ down to its simplest form.

Note that 14 and 7 are both divisible by 7:
- $14 \div 7 = 2$
- $7 \div 7 = 1$

This means that 7 is a common factor of the numerator and denominator. Divide the numerator and denominator both by 7 in order to reduce the fraction.

$$\frac{14}{7} = \frac{14 \div 7}{7 \div 7} = \frac{2}{1} = 2$$

The fraction $\frac{14}{7}$ reduces down to 2. We applied the rule that $\frac{2}{1}$ is equivalent to 2.

Date: _____   Score: ___/8   Name: _____

Chapter 2 Exercises – Part A

Directions: Reduce each fraction down to its simplest form.

❶

$$\frac{6}{9} =$$

❷

$$\frac{9}{12} =$$

❸

$$\frac{7}{14} =$$

❹

$$\frac{9}{1} =$$

❺

$$\frac{2}{6} =$$

❻

$$\frac{4}{2} =$$

❼

$$\frac{15}{25} =$$

❽

$$\frac{10}{12} =$$

❖ Check your answers using the Answer Key at the back of the book.

Date: _____   Score: ___/8   Name: _____

Chapter 2 Exercises – Part B

Directions: Reduce each fraction down to its simplest form.

9

$$\frac{8}{20} =$$

10

$$\frac{12}{3} =$$

11

$$\frac{24}{36} =$$

12

$$\frac{28}{36} =$$

13

$$\frac{10}{4} =$$

14

$$\frac{30}{5} =$$

15

$$\frac{20}{16} =$$

16

$$\frac{1}{1} =$$

❖ Check your answers using the Answer Key at the back of the book.

3 COMMON DENOMINATORS

In order to add, subtract, or compare two fractions, you must first find a common denominator. This chapter is focused on how to make a common denominator.

Consider the fractions $\frac{5}{2}$ and $\frac{4}{3}$. These fractions have different denominators:

- The fraction $\frac{5}{2}$ has a denominator of 2.
- The fraction $\frac{4}{3}$ has a denominator of 3.

The way to find a common denominator is to find a pair of fractions that are equivalent to the given fractions. We will do this by applying the following rule: If you multiply both the numerator and denominator of a fraction by the same number, you get an equivalent fraction.

Here is how we will find a common denominator for $\frac{5}{2}$ and $\frac{4}{3}$:

- We want to multiply the numerator and denominator of $\frac{5}{2}$ by a number in order to make a new fraction that is equivalent to $\frac{5}{2}$.
- We want to multiply the numerator and denominator of $\frac{4}{3}$ by a different number in order to make a new fraction that is equivalent to $\frac{4}{3}$.
- We will choose the number for each case such that the two new fractions (which are equivalent to $\frac{5}{2}$ and $\frac{4}{3}$) have the same denominator.

Definitions

- **numerator**: the number that appears in the top of a fraction.
- **denominator**: the number that appears in the bottom of a fraction.
- **common denominator**: when two fractions each have the same denominator.

Now we will find a common denominator for the fractions $\frac{5}{2}$ and $\frac{4}{3}$ using the strategy that we outlined on the previous page.

The denominators of the given fractions $\left(\frac{5}{2} \text{ and } \frac{4}{3}\right)$ are 2 and 3. What number is a multiple of both 2 and 3? The number 6 is the smallest multiple of both 2 and 3.

Therefore, we will make a common denominator of 6:

- Multiply the numerator and denominator of $\frac{5}{2}$ by 3 (because $2 \times 3 = 6$) in order to make an equivalent fraction with a denominator of 6.

$$\frac{5}{2} = \frac{5 \times 3}{2 \times 3} = \frac{15}{6}$$

- Multiply the numerator and denominator of $\frac{4}{3}$ by 2 (because $3 \times 2 = 6$) in order to make an equivalent fraction with a denominator of 6.

$$\frac{4}{3} = \frac{4 \times 2}{3 \times 2} = \frac{8}{6}$$

The fractions $\frac{15}{6}$ and $\frac{8}{6}$ are equivalent to the original fractions $\frac{5}{2}$ and $\frac{4}{3}$.

Since $\frac{15}{6}$ and $\frac{8}{6}$ have the same denominator, we have successfully made a common denominator from $\frac{5}{2}$ and $\frac{4}{3}$.

Furthermore, we found the lowest common denominator because 6 is the smallest multiple of the original denominators (2 and 3).

The examples in this chapter show you how to find the lowest common denominator. We will practice this valuable skill in this chapter, and then learn how to apply it in Chapters 4-6.

Definition

- **lowest common denominator**: when you use the least common multiple of the denominators of the given fractions to create a common denominator.

Finding Multiples

The multiples of a number are the products that you can make from multiplication with whole numbers. For example, the multiples of 5 are 5, 10, 15, 20, 25, 30, 35, 40, 45, 50, and so on. That's because $5 \times 1 = 5, 5 \times 2 = 10, 5 \times 3 = 15, 5 \times 4 = 20$, etc.

Following is a review of the first ten multiples of the numbers 2 through 10.
- 2, 4, 6, 8, 10, 12, 14, 16, 18, 20
- 3, 6, 9, 12, 15, 18, 21, 24, 27, 30
- 4, 8, 12, 16, 20, 24, 28, 32, 36, 40
- 5, 10, 15, 20, 25, 30, 35, 40, 45, 50
- 6, 12, 18, 24, 30, 36, 42, 48, 54, 60
- 7, 14, 21, 28, 35, 42, 49, 56, 63, 70
- 8, 16, 24, 32, 40, 48, 56, 64, 72, 80
- 9, 18, 27, 36, 45, 54, 63, 72, 81, 90
- 10, 20, 30, 40, 50, 60, 70, 80, 90, 100

Least Common Multiple

To find the lowest common denominator, we will use the least common multiple (as shown in the examples beginning on the following page).

Following are examples of least common multiples:
- The least common multiple of 3 and 5 is 15. This is formed from $3 \times 5 = 15$.
- The least common multiple of 4 and 7 is 28. This is formed from $4 \times 7 = 28$.
- The least common multiple of 6 and 8 is 24. That's because $6 \times 4 = 24$ and $8 \times 3 = 24$. Compare with the two previous examples: $6 \times 8 = 48$ isn't the least common multiple: It's 24, not 48. The reason for the difference is that 6 and 8 have a common factor of 2 (they are both even).
- The least common multiple of 18 and 27 is 54. That's because $18 \times 3 = 54$ and $27 \times 2 = 54$. Like the previous example, 18 and 27 have a common factor of 9, so the least common multiple is smaller than 18×27.
- The least common multiple of 5 and 10 is 10. That's because $5 \times 2 = 10$ and $10 \times 1 = 10$. This example is simpler because 10 is a multiple of 5.

Example: Express the fractions $\frac{1}{4}$ and $\frac{2}{5}$ with the lowest common denominator.

The least common multiple of 4 and 5 is the number 20:
- Multiply the numerator and denominator of $\frac{1}{4}$ by 5 (because $4 \times 5 = 20$) in order to make an equivalent fraction with a denominator of 20.
$$\frac{1}{4} = \frac{1 \times 5}{4 \times 5} = \frac{5}{20}$$
- Multiply the numerator and denominator of $\frac{2}{5}$ by 4 (because $5 \times 4 = 20$) in order to make an equivalent fraction with a denominator of 20.
$$\frac{2}{5} = \frac{2 \times 4}{5 \times 4} = \frac{8}{20}$$

The fractions $\frac{5}{20}$ and $\frac{8}{20}$ have the lowest common denominator from $\frac{1}{4}$ and $\frac{2}{5}$.

Example: Express the fractions $\frac{5}{6}$ and $\frac{3}{7}$ with the lowest common denominator.

The least common multiple of 6 and 7 is the number 42:
- Multiply the numerator and denominator of $\frac{5}{6}$ by 7 (because $6 \times 7 = 42$) in order to make an equivalent fraction with a denominator of 42.
$$\frac{5}{6} = \frac{5 \times 7}{6 \times 7} = \frac{35}{42}$$
- Multiply the numerator and denominator of $\frac{3}{7}$ by 6 (because $7 \times 6 = 42$) in order to make an equivalent fraction with a denominator of 42.
$$\frac{3}{7} = \frac{3 \times 6}{7 \times 6} = \frac{18}{42}$$

The fractions $\frac{35}{42}$ and $\frac{18}{42}$ have the lowest common denominator from $\frac{5}{6}$ and $\frac{3}{7}$.

Example: Express the fractions $\frac{4}{9}$ and $\frac{1}{6}$ with the lowest common denominator.

Long solution: The long way is to make a common multiple by multiplying 6 and 9:

- Multiply the numerator and denominator of $\frac{4}{9}$ by 6 (because $9 \times 6 = 54$) in order to make an equivalent fraction with a denominator of 54.

$$\frac{4}{9} = \frac{4 \times 6}{9 \times 6} = \frac{24}{54}$$

- Multiply the numerator and denominator of $\frac{1}{6}$ by 9 (because $6 \times 9 = 54$) in order to make an equivalent fraction with a denominator of 54.

$$\frac{1}{6} = \frac{1 \times 9}{6 \times 9} = \frac{9}{54}$$

Although the fractions $\frac{24}{54}$ and $\frac{9}{54}$ now have a common denominator, this isn't the lowest common denominator. Both fractions can be reduced (as in Chapter 2) in order to have a common denominator of 18 as follows:

$$\frac{24}{54} = \frac{24 \div 3}{54 \div 3} = \frac{8}{18}$$

$$\frac{9}{54} = \frac{9 \div 3}{54 \div 3} = \frac{3}{18}$$

The fractions $\frac{8}{18}$ and $\frac{3}{18}$ have the lowest common denominator from $\frac{4}{9}$ and $\frac{1}{6}$.

Short solution: The least common multiple of 9 and 6 is the number 18:

- Multiply the numerator and denominator of $\frac{4}{9}$ by 2 (because $9 \times 2 = 18$) in order to make an equivalent fraction with a denominator of 18.

$$\frac{4}{9} = \frac{4 \times 2}{9 \times 2} = \frac{8}{18}$$

- Multiply the numerator and denominator of $\frac{1}{6}$ by 3 (because $6 \times 3 = 18$) in order to make an equivalent fraction with a denominator of 18.

$$\frac{1}{6} = \frac{1 \times 3}{6 \times 3} = \frac{3}{18}$$

The fractions $\frac{8}{18}$ and $\frac{3}{18}$ have the lowest common denominator from $\frac{4}{9}$ and $\frac{1}{6}$.

Compare: 18 is the least common multiple of 9 and 6. Your solutions will be shorter if you can think of the smallest multiple that can be made from the two given denominators. (If not, that's okay, but you may need to complete more steps.)

The same method works to find the common denominator between a whole number and a fraction. For example, consider 2 and $\frac{5}{4}$. We can write the whole number over a denominator of one, applying the rule that $\frac{2}{1} = 2$. We learned this in Chapter 2 (see page 20): We can think of $\frac{2}{1}$ as being $2 \div 1$, which is equal to 2.

Example: Express 2 and the fraction $\frac{5}{4}$ with the lowest common denominator.

First rewrite the whole number 2 as a fraction by dividing by one: $2 = \frac{2}{1}$. Now we have two fractions, $\frac{2}{1}$ and $\frac{5}{4}$. The least common multiple of 1 and 4 is the number 4:

- Multiply the numerator and denominator of $\frac{2}{1}$ by 4 (because $1 \times 4 = 4$) in order to make an equivalent fraction with a denominator of 4.
$$\frac{2}{1} = \frac{2 \times 4}{1 \times 4} = \frac{8}{4}$$
- The fraction $\frac{5}{4}$ already has a denominator of 4. We don't need to change it.

The fractions $\frac{8}{4}$ and $\frac{5}{4}$ have the lowest common denominator from 2 and $\frac{5}{4}$.

Example: Express 3 and the fraction $\frac{2}{5}$ with the lowest common denominator.

First rewrite the whole number 3 as a fraction by dividing by one: $3 = \frac{3}{1}$. Now we have two fractions, $\frac{3}{1}$ and $\frac{2}{5}$. The least common multiple of 1 and 5 is the number 5:

- Multiply the numerator and denominator of $\frac{3}{1}$ by 5 (because $1 \times 5 = 5$) in order to make an equivalent fraction with a denominator of 5.
$$\frac{3}{1} = \frac{3 \times 5}{1 \times 5} = \frac{15}{5}$$
- The fraction $\frac{2}{5}$ already has a denominator of 5. We don't need to change it.

The fractions $\frac{15}{5}$ and $\frac{2}{5}$ have the lowest common denominator from 3 and $\frac{2}{5}$.

Date: _____   Score: ___/5   Name: _____

Chapter 3 Exercises – Part A

Directions: Express the fractions with the lowest common denominator.

❶

$\frac{1}{2}$ and $\frac{2}{3}$

❷

$\frac{5}{9}$ and $\frac{3}{4}$

❸

$\frac{1}{8}$ and $\frac{1}{12}$

❹

$\frac{7}{10}$ and $\frac{4}{15}$

❺

4 and $\frac{5}{3}$

❖ Check your answers using the Answer Key at the back of the book.

Date: _____   Score: ___/5   Name: _____

Chapter 3 Exercises – Part B

Directions: Express the fractions with the lowest common denominator.

❻

$\dfrac{7}{4}$ and $\dfrac{5}{3}$

❼

$\dfrac{11}{12}$ and $\dfrac{9}{16}$

❽

$\dfrac{5}{24}$ and $\dfrac{7}{36}$

❾

$\dfrac{9}{4}$ and 1

❿

$\dfrac{1}{2}$ and $\dfrac{7}{9}$

❖ Check your answers using the Answer Key at the back of the book.

Date: _____   Score: ___/5   Name: _____

Chapter 3 Exercises – Part C

Directions: Express the fractions with the lowest common denominator.

⑪

$\dfrac{3}{5}$ and $\dfrac{5}{3}$

⑫

$\dfrac{9}{25}$ and $\dfrac{11}{20}$

⑬

5 and $\dfrac{5}{2}$

⑭

$\dfrac{8}{15}$ and $\dfrac{10}{9}$

⑮

$\dfrac{3}{8}$ and $\dfrac{1}{4}$

❖ Check your answers using the Answer Key at the back of the book.

Date: _____   Score: ___/5   Name: _____

Chapter 3 Exercises – Part D

Directions: Express the fractions with the lowest common denominator.

⓰

$$\frac{8}{21} \text{ and } \frac{5}{28}$$

⓱

$$\frac{9}{8} \text{ and } \frac{8}{9}$$

⓲

$$\frac{3}{2} \text{ and } \frac{6}{11}$$

⓳

$$6 \text{ and } \frac{8}{3}$$

⓴

$$\frac{5}{12} \text{ and } \frac{5}{18}$$

❖ Check your answers using the Answer Key at the back of the book.

4 COMPARING FRACTIONS

Which fraction is bigger, $\frac{2}{3}$ or $\frac{5}{8}$? One way to determine this is to express each fraction with a common denominator using the method from Chapter 3.

The denominators of the given fractions $\left(\frac{2}{3} \text{ and } \frac{5}{8}\right)$ are 3 and 8. What number is a multiple of both 3 and 8? The number 24 is the smallest multiple of both 3 and 8.

Therefore, we will make a common denominator of 24:

- Multiply the numerator and denominator of $\frac{2}{3}$ by 8 (because $3 \times 8 = 24$) in order to make an equivalent fraction with a denominator of 24.

$$\frac{2}{3} = \frac{2 \times 8}{3 \times 8} = \frac{16}{24}$$

- Multiply the numerator and denominator of $\frac{5}{8}$ by 3 (because $8 \times 3 = 24$) in order to make an equivalent fraction with a denominator of 24.

$$\frac{5}{8} = \frac{5 \times 3}{8 \times 3} = \frac{15}{24}$$

The fractions $\frac{16}{24}$ and $\frac{15}{24}$ are equivalent to the original fractions $\frac{2}{3}$ and $\frac{5}{8}$. We can see that the fraction $\frac{16}{24}$ is greater than $\frac{15}{24}$ according to the following rule: If two fractions have the same denominator, the fraction with the greater numerator is the larger fraction. The fraction $\frac{2}{3}$ is greater than $\frac{5}{8}$ (since $\frac{16}{24} = \frac{2}{3}$ and $\frac{15}{24} = \frac{5}{8}$).

Definitions

- **numerator**: the number that appears in the top of a fraction.
- **denominator**: the number that appears in the bottom of a fraction.
- **common denominator**: when two fractions each have the same denominator.

We will use the following symbols in our answers in this chapter.

- The symbol $<$ means "is less than."
- The symbol $>$ means "is greater than."
- The symbol $=$ means "is equal to."

Following are some examples of these symbols:

- $3 < 5$
- $5 > 3$
- $4 = 4$

Example: Compare the fractions $\frac{5}{3}$ and $\frac{7}{4}$.

First we will find a common denominator using the method from Chapter 3. The least common multiple of 3 and 4 is the number 12:

- Multiply the numerator and denominator of $\frac{5}{3}$ by 4 (because $3 \times 4 = 12$) in order to make an equivalent fraction with a denominator of 12.

$$\frac{5}{3} = \frac{5 \times 4}{3 \times 4} = \frac{20}{12}$$

- Multiply the numerator and denominator of $\frac{7}{4}$ by 3 (because $4 \times 3 = 12$) in order to make an equivalent fraction with a denominator of 12.

$$\frac{7}{4} = \frac{7 \times 3}{4 \times 3} = \frac{21}{12}$$

Now that we have a common denominator of 12, it's easy to see that $\frac{20}{12} < \frac{21}{12}$ (when two fractions have the same denominator, the fraction with the greater numerator is larger). Since $\frac{20}{12} = \frac{5}{3}$ and $\frac{21}{12} = \frac{7}{4}$, it follows that $\frac{5}{3} < \frac{7}{4}$ (meaning that $\frac{5}{3}$ is less than $\frac{7}{4}$).

Example: Compare the fractions $\frac{2}{5}$ and $\frac{4}{11}$.

First we will find a common denominator using the method from Chapter 3. The least common multiple of 11 and 5 is the number 55:

- Multiply the numerator and denominator of $\frac{2}{5}$ by 11 (because $5 \times 11 = 55$) in order to make an equivalent fraction with a denominator of 55.

$$\frac{2}{5} = \frac{2 \times 11}{5 \times 11} = \frac{22}{55}$$

- Multiply the numerator and denominator of $\frac{4}{11}$ by 5 (because $11 \times 5 = 55$) in order to make an equivalent fraction with a denominator of 55.

$$\frac{4}{11} = \frac{4 \times 5}{11 \times 5} = \frac{20}{55}$$

Now that we have a common denominator of 55, it's easy to see that $\frac{22}{55} > \frac{20}{55}$. Since $\frac{22}{55} = \frac{2}{5}$ and $\frac{20}{55} = \frac{4}{11}$, it follows that $\frac{2}{5} > \frac{4}{11}$ (meaning that $\frac{2}{5}$ is greater than $\frac{4}{11}$).

Example: Compare the fractions $\frac{4}{6}$ and $\frac{6}{9}$.

First we will find a common denominator using the method from Chapter 3. The least common multiple of 6 and 9 is the number 18:

- Multiply the numerator and denominator of $\frac{4}{6}$ by 3 (because $6 \times 3 = 18$) in order to make an equivalent fraction with a denominator of 18.

$$\frac{4}{6} = \frac{4 \times 3}{6 \times 3} = \frac{12}{18}$$

- Multiply the numerator and denominator of $\frac{6}{9}$ by 2 (because $9 \times 2 = 18$) in order to make an equivalent fraction with a denominator of 18.

$$\frac{6}{9} = \frac{6 \times 2}{9 \times 2} = \frac{12}{18}$$

Now that we have a common denominator of 18, it's easy to see that $\frac{12}{18} = \frac{12}{18}$. Since $\frac{12}{18} = \frac{4}{6}$ and $\frac{12}{18} = \frac{6}{9}$, it follows that $\frac{4}{6} = \frac{6}{9}$ (meaning that $\frac{4}{6}$ is equal to $\frac{6}{9}$).

Alternate Method

There is a simpler method that you can use to compare two fractions. This method involves cross multiplying:
- Multiply the numerator of the left fraction with the denominator of the right fraction. Write this product on the left (it corresponds to the left fraction).
- Multiply the numerator of the right fraction with the denominator of the left fraction. Write this product on the right (it corresponds to the right fraction).

Whichever product is bigger, the corresponding fraction is bigger.

For example, consider the fractions $\frac{1}{3}$ and $\frac{3}{8}$.
- Cross multiplying, we get 1×8 and 3×3.
- This simplifies to 8 and 9.
- Since $8 < 9$, it follows that $\frac{1}{3} < \frac{3}{8}$ (meaning that $\frac{1}{3}$ is less than $\frac{3}{8}$).

If we solved the problem with the longer method, we would multiply the numerator and denominator of $\frac{1}{3}$ by 8 and the numerator and denominator of $\frac{3}{8}$ by 3.

$$\frac{1}{3} = \frac{1 \times 8}{3 \times 8} = \frac{8}{24} \text{ and } \frac{3}{8} = \frac{3 \times 3}{8 \times 3} = \frac{9}{24}$$

Since $\frac{8}{24} < \frac{9}{24}$, we obtain the same answer: $\frac{1}{3} < \frac{3}{8}$.

Why does cross multiplying work?

To see this, compare the two solutions above:
- With the cross multiplying method, we found 1×8 and 3×3.
- With the common denominator method, we also found 1×8 and 3×3.
- The denominators were the same: 3×8 and 8×3 both equal 24.
- Therefore, we only really need to compare the new numerators, which are 1×8 and 3×3.
- That's what the cross multiplying method does. It effectively just looks at the new numerators, which you get by cross multiplying.

The following examples illustrate the cross multiplying method.

Example: Compare the fractions $\frac{9}{2}$ and $\frac{30}{7}$.

Cross multiply as follows:
- Multiply the left numerator (9) by the right denominator (7). Write this on the left.
- Multiply the right numerator (30) by the left denominator (2). Write this on the right.
- We get 9×7 and 30×2.
- This simplifies to 63 and 60.

Since $63 > 60$, it follows that $\frac{9}{2} > \frac{30}{7}$ (meaning that $\frac{9}{2}$ is greater than $\frac{30}{7}$).

Example: Compare the fractions $\frac{7}{16}$ and $\frac{1}{2}$.

Cross multiply as follows:
- Multiply the left numerator (7) by the right denominator (2). Write this on the left.
- Multiply the right numerator (1) by the left denominator (16). Write this on the right.
- We get 7×2 and 1×16.
- This simplifies to 14 and 16.

Since $14 < 16$, it follows that $\frac{7}{16} < \frac{1}{2}$ (meaning that $\frac{7}{16}$ is less than $\frac{1}{2}$).

Note: Both methods have merit.
- The common denominator method offers practice with the skill of finding a common denominator. This skill is used for adding and subtracting fractions (as we will see in Chapters 5-6).
- The cross multiplying method provides early exposure to an important skill that is taught in algebra courses, so it is a good preparation skill. Students like to use this method for comparing fractions because it makes the solution shorter.

You can use the same methods to compare a whole number to a fraction. For example, consider 2 and $\frac{8}{3}$. We can write the whole number over a denominator of one, applying the rule that $\frac{2}{1} = 2$. We learned this in Chapter 2 (see page 20): We can think of $\frac{2}{1}$ as being $2 \div 1$, which is equal to 2.

Example: Compare the whole number 2 to the fraction $\frac{8}{3}$.

First rewrite the whole number 2 as a fraction by dividing by one: $2 = \frac{2}{1}$. Now we have two fractions, $\frac{2}{1}$ and $\frac{8}{3}$. Cross multiply as follows:

- Multiply the left numerator (2) by the right denominator (3). Write this on the left.
- Multiply the right numerator (8) by the left denominator (1). Write this on the right.
- We get 2×3 and 8×1.
- This simplifies to 6 and 8.

Since $6 < 8$, it follows that $\frac{2}{1} < \frac{8}{3}$ (meaning that 2 is less than $\frac{8}{3}$).

Example: Compare the fraction $\frac{14}{2}$ to the whole number 7.

First rewrite the whole number 7 as a fraction by dividing by one: $7 = \frac{7}{1}$. Now we have two fractions, $\frac{14}{2}$ and $\frac{7}{1}$. Cross multiply as follows:

- Multiply the left numerator (14) by the right denominator (1). Write this on the left.
- Multiply the right numerator (7) by the left denominator (2). Write this on the right.
- We get 14×1 and 7×2.
- This simplifies to 14 and 14.

Since $14 = 14$, it follows that $\frac{14}{2} = \frac{7}{1}$ (meaning that $\frac{14}{2}$ is equal to 7). This particular example is much simpler if you realize that $\frac{14}{2}$ equals $14 \div 2 = 7$.

Date: _____   Score: ___/5   Name: _____

Chapter 4 Exercises – Part A

Directions: Compare the fractions. Then write <, >, or = in the dashed box.

❶

$$\frac{1}{2} \; \boxed{} \; \frac{3}{7}$$

❷

$$\frac{6}{5} \; \boxed{} \; \frac{3}{2}$$

❸

$$\frac{7}{9} \; \boxed{} \; \frac{5}{8}$$

❹

$$\frac{7}{3} \; \boxed{} \; 2$$

❺

$$\frac{3}{4} \; \boxed{} \; \frac{12}{16}$$

❖ Check your answers using the Answer Key at the back of the book.

Date: _____   Score: ___/5   Name: _____

Chapter 4 Exercises – Part B

Directions: Compare the fractions. Then write $<$, $>$, or $=$ in the dashed box.

❻

$$\frac{2}{7} \ \boxed{} \ \frac{3}{8}$$

❼

$$\frac{4}{3} \ \boxed{} \ \frac{5}{4}$$

❽

$$\frac{5}{6} \ \boxed{} \ \frac{7}{9}$$

❾

$$\frac{5}{2} \ \boxed{} \ \frac{15}{6}$$

❿

$$1 \ \boxed{} \ \frac{10}{9}$$

❖ Check your answers using the Answer Key at the back of the book.

Date: _____   Score: ___/5   Name: _____

Chapter 4 Exercises – Part C

Directions: Compare the fractions. Then write $<$, $>$, or $=$ in the dashed box.

⑪

$\dfrac{3}{5}$ ⬚ $\dfrac{7}{9}$

⑫

$\dfrac{1}{6}$ ⬚ $\dfrac{1}{8}$

⑬

$\dfrac{3}{4}$ ⬚ $\dfrac{8}{9}$

⑭

$\dfrac{6}{5}$ ⬚ $\dfrac{4}{3}$

⑮

$\dfrac{12}{4}$ ⬚ 3

❖ Check your answers using the Answer Key at the back of the book.

Date: _____   Score: ___/5   Name: _____

Chapter 4 Exercises – Part D

Directions: Compare the fractions. Then write <, >, or = in the dashed box.

⑯

$$\frac{19}{6} \quad \boxed{} \quad \frac{22}{7}$$

⑰

$$\frac{11}{3} \quad \boxed{} \quad 4$$

⑱

$$\frac{10}{8} \quad \boxed{} \quad \frac{15}{12}$$

⑲

$$\frac{18}{5} \quad \boxed{} \quad \frac{18}{7}$$

⑳

$$\frac{1}{4} \quad \boxed{} \quad \frac{3}{16}$$

❖ Check your answers using the Answer Key at the back of the book.

5 ADDING FRACTIONS

The way to add two fractions together is to first make a common denominator using the method from Chapter 3. Once you make a common denominator, you may then add the numerators together, keeping the common denominator unchanged.

For example, consider the two fractions added together below.

$$\frac{3}{8} + \frac{4}{3} = \frac{3 \times 3}{8 \times 3} + \frac{4 \times 8}{3 \times 8} = \frac{9}{24} + \frac{32}{24} = \frac{9 + 32}{24} = \frac{41}{24}$$

In the first step, we multiplied the numerator and denominator of $\frac{3}{8}$ by 3 and we multiplied the numerator and denominator of $\frac{4}{3}$ by 8 in order to create a common denominator of 24 (since the least common multiple of 8 and 3 is the number 24): $8 \times 3 = 24$ and $3 \times 8 = 24$. This way, we found two new fractions, $\frac{9}{24}$ and $\frac{32}{24}$, with a common denominator. We then added the numerators (9 and 32) of these equivalent fractions together to get our final answer of $\frac{41}{24}$. You can find additional examples broken down into steps beginning on the following page.

Notes: Don't add the numerators together until you make a common denominator. Never add the denominators together: Only add numerators (after making a common denominator).

You may need to reduce your answer (as discussed in Chapter 2), especially if you don't use the lowest common denominator (Chapter 3 described how to find the lowest common denominator).

Definitions

- **numerator**: the number that appears in the top of a fraction.
- **denominator**: the number that appears in the bottom of a fraction.
- **common denominator**: when two fractions each have the same denominator.

Example: Add the fractions $\frac{2}{5}$ and $\frac{3}{4}$.

Step by step explanation: First we will find a common denominator using the method from Chapter 3. The least common multiple of 5 and 4 is the number 20:

- Multiply the numerator and denominator of $\frac{2}{5}$ by 4 (because $5 \times 4 = 20$) in order to make an equivalent fraction with a denominator of 20.
$$\frac{2}{5} = \frac{2 \times 4}{5 \times 4} = \frac{8}{20}$$

- Multiply the numerator and denominator of $\frac{3}{4}$ by 5 (because $4 \times 5 = 20$) in order to make an equivalent fraction with a denominator of 20.
$$\frac{3}{4} = \frac{3 \times 5}{4 \times 5} = \frac{15}{20}$$

- Now we can express the problem with common denominators $\left(\frac{8}{20} \text{ and } \frac{15}{20}\right)$, since $\frac{2}{5} = \frac{8}{20}$ and $\frac{3}{4} = \frac{15}{20}$.
$$\frac{2}{5} + \frac{3}{4} = \frac{8}{20} + \frac{15}{20}$$

- Once we have two equivalent fractions with a common denominator, we may add the numerators together, leaving the common denominator unchanged.
$$\frac{8}{20} + \frac{15}{20} = \frac{8 + 15}{20} = \frac{23}{20}$$

Short solution: We could organize all of these steps into a single line as follows.
$$\frac{2}{5} + \frac{3}{4} = \frac{2 \times 4}{5 \times 4} + \frac{3 \times 5}{4 \times 5} = \frac{8}{20} + \frac{15}{20} = \frac{8 + 15}{20} = \frac{23}{20}$$
Here is a summary of what we did:

- We made a common denominator. The least common multiple of 5 and 4 is the number 20.
- We multiplied $\frac{2}{5}$ by $\frac{4}{4}$ since $5 \times 4 = 20$. We multiplied $\frac{3}{4}$ by $\frac{5}{5}$ since $4 \times 5 = 20$. This gave us two new fractions $\left(\frac{8}{20} \text{ and } \frac{15}{20}\right)$ with a denominator of 20.
- Once the denominator was common, we added the two numerators together: $8 + 15 = 23$.

Our final answer is $\frac{23}{20}$.

Example: Add the fractions $\frac{5}{8}$ and $\frac{7}{6}$.

Step by step explanation: First we will find a common denominator using the method from Chapter 3. The least common multiple of 8 and 6 is the number 24:

- **Note**: If you multiply $8 \times 6 = 48$ to make your common denominator, in this example it won't be the lowest common denominator – so you would need to reduce your answer at the end (like we did in Chapter 2). The solution is somewhat simpler if you realize that 24 is a multiple of both 8 and 6 (because $8 \times 3 = 24$ and $6 \times 4 = 24$). In this example, 24 is the lowest common denominator.

- Multiply the numerator and denominator of $\frac{5}{8}$ by 3 (because $8 \times 3 = 24$) in order to make an equivalent fraction with a denominator of 24.

$$\frac{5}{8} = \frac{5 \times 3}{8 \times 3} = \frac{15}{24}$$

- Multiply the numerator and denominator of $\frac{7}{6}$ by 4 (because $6 \times 4 = 24$) in order to make an equivalent fraction with a denominator of 24.

$$\frac{7}{6} = \frac{7 \times 4}{6 \times 4} = \frac{28}{24}$$

- Now we can express the problem with common denominators $\left(\frac{15}{24} \text{ and } \frac{28}{24}\right)$, since $\frac{5}{8} = \frac{15}{24}$ and $\frac{7}{6} = \frac{28}{24}$.

$$\frac{5}{8} + \frac{7}{6} = \frac{15}{24} + \frac{28}{24}$$

- Once we have two equivalent fractions with a common denominator, we may add the numerators together, leaving the common denominator unchanged.

$$\frac{15}{24} + \frac{28}{24} = \frac{15 + 28}{24} = \frac{43}{24}$$

Short solution: We could organize all of these steps into a single line as follows.

$$\frac{5}{8} + \frac{7}{6} = \frac{5 \times 3}{8 \times 3} + \frac{7 \times 4}{6 \times 4} = \frac{15}{24} + \frac{28}{24} = \frac{15 + 28}{24} = \frac{43}{24}$$

Our final answer is $\frac{43}{24}$.

You can use the same method to add a whole number to a fraction. For example, consider adding 2 to $\frac{4}{3}$. We can write the whole number over a denominator of one, applying the rule that $\frac{2}{1} = 2$. We learned this in Chapter 2 (see page 20): We can think of $\frac{2}{1}$ as being $2 \div 1$, which is equal to 2.

Example: Add the whole number 2 to the fraction $\frac{4}{3}$.

Step by step explanation:

- First rewrite the whole number 2 as a fraction by dividing by one: $2 = \frac{2}{1}$.
- When adding a whole number to a reduced fraction, the lowest common denominator is automatically the denominator of the fraction. In this example, the lowest common denominator is 3 (the denominator of $\frac{4}{3}$).
- Multiply the numerator and denominator of $\frac{2}{1}$ by 3 (because $1 \times 3 = 3$) in order to make an equivalent fraction with a denominator of 3.
$$\frac{2}{1} = \frac{2 \times 3}{1 \times 3} = \frac{6}{3}$$
- The given fraction $\frac{4}{3}$ already has the same denominator. We don't need to change it.
- Now we can express the problem with common denominators $\left(\frac{6}{3} \text{ and } \frac{4}{3}\right)$, since $2 = \frac{2}{1} = \frac{6}{3}$.
$$2 + \frac{4}{3} = \frac{2}{1} + \frac{4}{3} = \frac{6}{3} + \frac{4}{3}$$
- Once we have two equivalent fractions with a common denominator, we may add the numerators together, leaving the common denominator unchanged.
$$\frac{6}{3} + \frac{4}{3} = \frac{6 + 4}{3} = \frac{10}{3}$$

Short solution: We could organize all of these steps into a single line as follows.
$$2 + \frac{4}{3} = \frac{2}{1} + \frac{4}{3} = \frac{2 \times 3}{1 \times 3} + \frac{4}{3} = \frac{6}{3} + \frac{4}{3} = \frac{6 + 4}{3} = \frac{10}{3}$$
Our final answer is $\frac{10}{3}$.

Date: _____   Score: ___/5   Name: _____

Chapter 5 Exercises – Part A

Directions: Add the fractions by first finding a common denominator.

❶

$$\frac{1}{3} + \frac{1}{4} =$$

❷

$$\frac{2}{3} + \frac{4}{5} =$$

❸

$$\frac{1}{4} + \frac{1}{8} =$$

❹

$$\frac{3}{7} + \frac{5}{2} =$$

❺

$$\frac{9}{4} + 3 =$$

❖ Check your answers using the Answer Key at the back of the book.

Date: _____   Score: ___/5   Name: _____

Chapter 5 Exercises – Part B

Directions: Add the fractions by first finding a common denominator.

❻

$$\frac{5}{6} + \frac{4}{9} =$$

❼

$$\frac{8}{3} + \frac{8}{11} =$$

❽

$$\frac{15}{4} + \frac{10}{3} =$$

❾

$$1 + \frac{7}{5} =$$

❿

$$\frac{7}{11} + \frac{4}{11} =$$

❖Check your answers using the Answer Key at the back of the book.

Date: _____   Score: ___/5   Name: _____

Chapter 5 Exercises – Part C

Directions: Add the fractions by first finding a common denominator.

⑪

$$\frac{2}{9} + \frac{5}{12} =$$

⑫

$$\frac{4}{3} + \frac{3}{4} =$$

⑬

$$\frac{1}{2} + \frac{2}{3} =$$

⑭

$$\frac{7}{6} + \frac{6}{5} =$$

⑮

$$\frac{15}{4} + 4 =$$

❖ Check your answers using the Answer Key at the back of the book.

Date: _____   Score: ___/5   Name: _____

Chapter 5 Exercises – Part D

Directions: Add the fractions by first finding a common denominator.

⑯

$$\frac{7}{8} + \frac{8}{9} =$$

⑰

$$\frac{1}{16} + \frac{5}{8} =$$

⑱

$$\frac{4}{5} + \frac{6}{7} =$$

⑲

$$5 + \frac{20}{7} =$$

⑳

$$\frac{1}{4} + \frac{1}{4} =$$

❖ Check your answers using the Answer Key at the back of the book.

6 SUBTRACTING FRACTIONS

Subtracting fractions is very similar to adding fractions (Chapter 5). You first make a common denominator using the method from Chapter 3. Once you make a common denominator, you may then subtract the numerators of the fractions, keeping the common denominator unchanged.

For example, consider the two fractions subtracted below.

$$\frac{5}{4} - \frac{2}{3} = \frac{5 \times 3}{4 \times 3} - \frac{2 \times 4}{3 \times 4} = \frac{15}{12} - \frac{8}{12} = \frac{15 - 8}{12} = \frac{7}{12}$$

In the first step, we multiplied the numerator and denominator of $\frac{5}{4}$ by 3 and we multiplied the numerator and denominator of $\frac{2}{3}$ by 4 in order to create a common denominator of 12 (since the least common multiple of 4 and 3 is the number 12): $4 \times 3 = 12$ and $3 \times 4 = 12$. This way, we found two new fractions, $\frac{15}{12}$ and $\frac{8}{12}$, with a common denominator. We then subtracted the numerators (15 and 8) of these equivalent fractions to get our final answer of $\frac{7}{12}$. You can find additional examples broken down into steps beginning on the following page.

Note: You may need to reduce your answer (as discussed in Chapter 2). This may be necessary even if you use the lowest common denominator.

Definitions

- **numerator**: the number that appears in the top of a fraction.
- **denominator**: the number that appears in the bottom of a fraction.
- **common denominator**: when two fractions each have the same denominator.

Example: Perform the subtraction $\frac{4}{5} - \frac{1}{2}$.

Step by step explanation: First we will find a common denominator using the method from Chapter 3. The least common multiple of 5 and 2 is the number 10:

- Multiply the numerator and denominator of $\frac{4}{5}$ by 2 (because $5 \times 2 = 10$) in order to make an equivalent fraction with a denominator of 10.

$$\frac{4}{5} = \frac{4 \times 2}{5 \times 2} = \frac{8}{10}$$

- Multiply the numerator and denominator of $\frac{1}{2}$ by 5 (because $2 \times 5 = 10$) in order to make an equivalent fraction with a denominator of 10.

$$\frac{1}{2} = \frac{1 \times 5}{2 \times 5} = \frac{5}{10}$$

- Now we can express the problem with common denominators $\left(\frac{8}{10} \text{ and } \frac{5}{10}\right)$, since $\frac{4}{5} = \frac{8}{10}$ and $\frac{1}{2} = \frac{5}{10}$.

$$\frac{4}{5} - \frac{1}{2} = \frac{8}{10} - \frac{5}{10}$$

- Once we have two equivalent fractions with a common denominator, we may subtract the numerators, leaving the common denominator unchanged.

$$\frac{8}{10} - \frac{5}{10} = \frac{8-5}{10} = \frac{3}{10}$$

Short solution: We could organize all of these steps into a single line as follows.

$$\frac{4}{5} - \frac{1}{2} = \frac{4 \times 2}{5 \times 2} - \frac{1 \times 5}{2 \times 5} = \frac{8}{10} - \frac{5}{10} = \frac{8-5}{10} = \frac{3}{10}$$

Here is a summary of what we did:

- We made a common denominator. The least common multiple of 5 and 2 is the number 10.
- We multiplied $\frac{4}{5}$ by $\frac{2}{2}$ since $5 \times 2 = 10$. We multiplied $\frac{1}{2}$ by $\frac{5}{5}$ since $2 \times 5 = 10$. This gave us two new fractions $\left(\frac{8}{10} \text{ and } \frac{5}{10}\right)$ with a denominator of 10.
- Once the denominator was common, we subtracted the numerators: $8 - 5 = 3$.

Our final answer is $\frac{3}{10}$.

Example: Perform the subtraction $\frac{4}{3} - \frac{1}{12}$.

Step by step explanation: First we will find a common denominator using the method from Chapter 3. The least common multiple of 3 and 12 is the number 12:

- Multiply the numerator and denominator of $\frac{4}{3}$ by 4 (because $3 \times 4 = 12$) in order to make an equivalent fraction with a denominator of 12.

$$\frac{4}{3} = \frac{4 \times 4}{3 \times 4} = \frac{16}{12}$$

- The given fraction $\frac{1}{12}$ already has the same denominator. We don't need to change it.

- Now we can express the problem with common denominators $\left(\frac{16}{12} \text{ and } \frac{1}{12}\right)$, since $\frac{4}{3} = \frac{16}{12}$.

$$\frac{4}{3} - \frac{1}{12} = \frac{16}{12} - \frac{1}{12}$$

- Once we have two equivalent fractions with a common denominator, we may subtract the numerators, leaving the common denominator unchanged.

$$\frac{16}{12} - \frac{1}{12} = \frac{16 - 1}{12} = \frac{15}{12}$$

- We're not finished yet because $\frac{15}{12}$ can be reduced. It is standard form to express your answer as a reduced fraction. Reduce $\frac{15}{12}$ using the method from Chapter 2. Note that 15 and 12 are both divisible by 3. Divide the numerator and denominator both by 3 in order to reduce the fraction.

$$\frac{15}{12} = \frac{15 \div 3}{12 \div 3} = \frac{5}{4}$$

Short solution: We could organize all of these steps into two lines as follows.

$$\frac{4}{3} - \frac{1}{12} = \frac{4 \times 4}{3 \times 4} - \frac{1}{12} = \frac{16}{12} - \frac{1}{12} = \frac{16 - 1}{12} = \frac{15}{12}$$

Since 15 and 12 are both divisible by 3, reduce the fraction $\frac{15}{12}$ by dividing the numerator and denominator both by 3.

$$\frac{15}{12} = \frac{15 \div 3}{12 \div 3} = \frac{5}{4}$$

Our final answer is $\frac{5}{4}$.

You can use the same method to subtract a fraction from a whole number, or to subtract a whole number from a fraction. For example, consider $6 - \frac{9}{2}$. We can write the whole number over a denominator of one, applying the rule that $\frac{6}{1} = 6$. We learned this in Chapter 2 (see page 20): We can think of $\frac{6}{1}$ as being $6 \div 1$, which is equal to 6.

Example: Perform the subtraction $6 - \frac{9}{2}$.

Step by step explanation:

- First rewrite the whole number 6 as a fraction by dividing by one: $6 = \frac{6}{1}$.

- When subtracting with a whole number and a reduced fraction, the lowest common denominator is automatically the denominator of the fraction. In this example, the lowest common denominator is 2 (the denominator of $\frac{9}{2}$).

- Multiply the numerator and denominator of $\frac{6}{1}$ by 2 (because $1 \times 2 = 2$) in order to make an equivalent fraction with a denominator of 2.
$$\frac{6}{1} = \frac{6 \times 2}{1 \times 2} = \frac{12}{2}$$

- The given fraction $\frac{9}{2}$ already has the same denominator. We don't need to change it.

- Now we can express the problem with common denominators $\left(\frac{12}{2} \text{ and } \frac{9}{2}\right)$, since $6 = \frac{6}{1} = \frac{12}{2}$.
$$6 - \frac{9}{2} = \frac{6}{1} - \frac{9}{2} = \frac{12}{2} - \frac{9}{2}$$

- Once we have two equivalent fractions with a common denominator, we may subtract the numerators, leaving the common denominator unchanged.
$$\frac{12}{2} - \frac{9}{2} = \frac{12-9}{2} = \frac{3}{2}$$

Short solution: We could organize all of these steps into a single line as follows.
$$6 - \frac{9}{2} = \frac{6}{1} - \frac{9}{2} = \frac{6 \times 2}{1 \times 2} - \frac{9}{2} = \frac{12}{2} - \frac{9}{2} = \frac{12-9}{2} = \frac{3}{2}$$
Our final answer is $\frac{3}{2}$.

Date: _____   Score: ___/5   Name: _____

Chapter 6 Exercises – Part A

Directions: Subtract the fractions by first finding a common denominator.

❶

$$\frac{1}{3} - \frac{1}{4} =$$

❷

$$\frac{3}{4} - \frac{2}{5} =$$

❸

$$\frac{7}{4} - \frac{5}{6} =$$

❹

$$\frac{1}{2} - \frac{1}{6} =$$

❺

$$\frac{17}{4} - 3 =$$

❖ Check your answers using the Answer Key at the back of the book.

Date: _____　　　Score: ___/5　　　Name: _____

Chapter 6 Exercises – Part B

Directions: Subtract the fractions by first finding a common denominator.

❻

$$\frac{4}{9} - \frac{3}{8} =$$

❼

$$\frac{5}{6} - \frac{5}{8} =$$

❽

$$\frac{9}{4} - \frac{5}{3} =$$

❾

$$1 - \frac{3}{5} =$$

❿

$$\frac{11}{5} - \frac{12}{9} =$$

❖ Check your answers using the Answer Key at the back of the book.

Date: _____   Score: ___/5   Name: _____

Chapter 6 Exercises – Part C

Directions: Subtract the fractions by first finding a common denominator.

⑪

$$\frac{5}{12} - \frac{7}{18} =$$

⑫

$$\frac{9}{20} - \frac{4}{15} =$$

⑬

$$2 - \frac{5}{4} =$$

⑭

$$\frac{8}{9} - \frac{2}{9} =$$

⑮

$$\frac{5}{4} - \frac{4}{5} =$$

❖ Check your answers using the Answer Key at the back of the book.

Date: _____   Score: ___/5   Name: _____

Chapter 6 Exercises – Part D

Directions: Subtract the fractions by first finding a common denominator.

⓰

$$\frac{15}{2} - \frac{8}{5} =$$

⓱

$$\frac{11}{24} - \frac{1}{36} =$$

⓲

$$\frac{16}{5} - \frac{3}{4} =$$

⓳

$$\frac{11}{18} - \frac{2}{9} =$$

⓴

$$3 - \frac{9}{4} =$$

❖ Check your answers using the Answer Key at the back of the book.

7 MULTIPLYING FRACTIONS

It is easier to multiply fractions than it is to add or subtract them. Why? When you multiply fractions, you don't need to find a common denominator.

The strategy for multiplying two fractions is simple:
- Multiply the two numerators together to make the new numerator.
- Multiply the two denominators together to make the new denominator.
- If the answer can be reduced, apply the method from Chapter 2 in order to reduce the fraction.

For example, consider the two fractions multiplied together below.

$$\frac{2}{3} \times \frac{5}{4} = \frac{2 \times 5}{3 \times 4} = \frac{10}{12}$$

The intermediate answer $\left(\frac{10}{12}\right)$ can be reduced because 10 and 12 are both divisible by 2. Applying the method from Chapter 2, we will divide both the numerator and denominator by 2 in order to reduce $\frac{10}{12}$.

$$\frac{10}{12} = \frac{10 \div 2}{12 \div 2} = \frac{5}{6}$$

The final answer is $\frac{5}{6}$.

The answer to a multiplication problem is called the product. For example, when we multiplied $\frac{2}{3} \times \frac{5}{4}$ in the above example, the product was $\frac{5}{6}$.

Definitions

- **numerator**: the number that appears in the top of a fraction.
- **denominator**: the number that appears in the bottom of a fraction.
- **product**: the number that results from multiplying two other numbers together. For example, in $3 \times 5 = 15$, the number 15 is called the product.

Example: Multiply the fractions $\frac{5}{6}$ and $\frac{3}{7}$.

- Multiply the numerators (5 and 3) together to form the new numerator.
- Multiply the denominators (6 and 7) together to form the new denominator.

$$\frac{5}{6} \times \frac{3}{7} = \frac{5 \times 3}{6 \times 7} = \frac{15}{42}$$

- The answer can be reduced since 15 and 42 are both divisible by 3. Divide the numerator and denominator both by 3.

$$\frac{15}{42} = \frac{15 \div 3}{42 \div 3} = \frac{5}{14}$$

Our final answer is $\frac{5}{14}$.

Example: Multiply the fractions $\frac{1}{4}$ and $\frac{5}{3}$.

- Multiply the numerators (1 and 5) together to form the new numerator.
- Multiply the denominators (4 and 3) together to form the new denominator.

$$\frac{1}{4} \times \frac{5}{3} = \frac{1 \times 5}{4 \times 3} = \frac{5}{12}$$

- This answer can't be reduced: 5 and 12 don't have any common factors.

Our final answer is $\frac{5}{12}$.

Example: Multiply the fractions $\frac{4}{7}$ and $\frac{7}{9}$.

- Multiply the numerators (4 and 7) together to form the new numerator.
- Multiply the denominators (7 and 9) together to form the new denominator.

$$\frac{4}{7} \times \frac{7}{9} = \frac{4 \times 7}{7 \times 9} = \frac{28}{63}$$

- The answer can be reduced since 28 and 63 are both divisible by 7. Divide the numerator and denominator both by 7.

$$\frac{28}{63} = \frac{28 \div 7}{63 \div 7} = \frac{4}{9}$$

Our final answer is $\frac{4}{9}$. (An alternate method is to cancel the 7's, noting that $\frac{7}{7} = 1$ in

$\frac{4}{7} \times \frac{7}{9} = \frac{4 \times 7}{7 \times 9} = \frac{4}{9} \times \frac{7}{7} = \frac{4}{9} \times 1 = \frac{4}{9}$.)

You can use the same method to multiply a whole number and a fraction together. For example, consider multiplying 6 by $\frac{2}{3}$. We can write the whole number over a denominator of one, applying the rule that $\frac{6}{1} = 6$. We learned this in Chapter 2 (see page 20): We can think of $\frac{6}{1}$ as being $6 \div 1$, which is equal to 6.

Example: Multiply the whole number 6 by the fraction $\frac{2}{3}$.

First rewrite the whole number 6 as a fraction by dividing by one: $6 = \frac{6}{1}$. Now we have two fractions, $\frac{6}{1}$ and $\frac{2}{3}$.
- Multiply the numerators (6 and 2) together to form the new numerator.
- Multiply the denominators (1 and 3) together to form the new denominator.

$$\frac{6}{1} \times \frac{2}{3} = \frac{6 \times 2}{1 \times 3} = \frac{12}{3}$$

- The answer can be reduced since 12 and 3 are both divisible by 3. Divide the numerator and denominator both by 3.

$$\frac{12}{3} = \frac{12 \div 3}{3 \div 3} = \frac{4}{1} = 4$$

Our final answer is 4. Note that $\frac{4}{1} = 4 \div 1 = 4$. Alternatively, go back a step and note that $\frac{12}{3} = 12 \div 3 = 4$ (if necessary, review page 20).

Example: Multiply the fraction $\frac{4}{5}$ by the whole number 8.

First rewrite the whole number 8 as a fraction by dividing by one: $8 = \frac{8}{1}$. Now we have two fractions, $\frac{4}{5}$ and $\frac{8}{1}$.
- Multiply the numerators (4 and 8) together to form the new numerator.
- Multiply the denominators (5 and 1) together to form the new denominator.

$$\frac{4}{5} \times \frac{8}{1} = \frac{4 \times 8}{5 \times 1} = \frac{32}{5}$$

- This answer can't be reduced: 32 and 5 don't have any common factors.

Our final answer is $\frac{32}{5}$.

Date: _____   Score: ___/5   Name: _____

Chapter 7 Exercises – Part A

Directions: Multiply the given fractions. Express your answer as a reduced fraction.

❶

$$\frac{3}{2} \times \frac{1}{5} =$$

❷

$$\frac{4}{9} \times \frac{5}{2} =$$

❸

$$\frac{5}{6} \times \frac{3}{8} =$$

❹

$$\frac{1}{8} \times \frac{1}{2} =$$

❺

$$\frac{4}{3} \times 4 =$$

❖ Check your answers using the Answer Key at the back of the book.

Date: _____   Score: ___/5   Name: _____

Chapter 7 Exercises – Part B

Directions: Multiply the given fractions. Express your answer as a reduced fraction.

❻

$$\frac{9}{5} \times \frac{4}{3} =$$

❼

$$\frac{2}{5} \times \frac{4}{7} =$$

❽

$$\frac{2}{5} \times \frac{15}{4} =$$

❾

$$1 \times \frac{7}{5} =$$

❿

$$\frac{9}{16} \times \frac{8}{9} =$$

❖ Check your answers using the Answer Key at the back of the book.

Date: _____ Score: ___/5 Name: _____

Chapter 7 Exercises – Part C

Directions: Multiply the given fractions. Express your answer as a reduced fraction.

⓫

$$\frac{12}{5} \times \frac{1}{18} =$$

⓬

$$\frac{2}{3} \times \frac{2}{3} =$$

⓭

$$\frac{15}{3} \times \frac{9}{20} =$$

⓮

$$\frac{7}{4} \times \frac{4}{7} =$$

⓯

$$\frac{5}{6} \times 4 =$$

❖ Check your answers using the Answer Key at the back of the book.

8 RECIPROCALS

To find the reciprocal of a fraction, simply swap the numerator and denominator.

For example, the reciprocal of $\frac{3}{5}$ is $\frac{5}{3}$ and the reciprocal of $\frac{7}{4}$ is $\frac{4}{7}$.

The reciprocal of a fraction is a multiplicative inverse. This means that when you multiply a fraction and its reciprocal together, the answer equals one. For example, when you multiply $\frac{3}{5}$ by $\frac{5}{3}$, you get $\frac{3}{5} \times \frac{5}{3} = \frac{3 \times 5}{5 \times 3} = \frac{15}{15} = 1$. (The 3 and 5 both cancel.)

To find the reciprocal of a whole number like 8, first rewrite 8 as $\frac{8}{1}$ (see page 20), and then swap the numerator and denominator. The reciprocal of 8 is $\frac{1}{8}$.

The reciprocal of any whole number equals one divided by the whole number. For example, the reciprocal of 15 is $\frac{1}{15}$ and the reciprocal of 100 is $\frac{1}{100}$.

Recall that any number divided by one equals itself. For example, $7 \div 1 = 7$, which can be expressed as $\frac{7}{1} = 7$. We can apply this to reason that the reciprocal of $\frac{1}{7}$ is 7.

Reciprocals come in pairs. For example, the reciprocal of $\frac{5}{12}$ is $\frac{12}{5}$ and the reciprocal of $\frac{12}{5}$ is $\frac{5}{12}$. Similarly, the reciprocal 11 is $\frac{1}{11}$ and the reciprocal of $\frac{1}{11}$ is 11.

Definitions

- **numerator**: the number that appears in the top of a fraction.
- **denominator**: the number that appears in the bottom of a fraction.
- **reciprocal**: what you get when you swap the numerator and denominator.
- **multiplicative inverse**: the reciprocal of a fraction.

Example: Find the reciprocal of $\frac{3}{7}$.

Swap the numerator (3) with the denominator (7). The reciprocal is $\frac{7}{3}$.

Example: Find the reciprocal of $\frac{19}{20}$.

Swap the numerator (19) with the denominator (20). The reciprocal is $\frac{20}{19}$.

Example: Find the reciprocal of $\frac{1}{6}$.

Swap the numerator (1) with the denominator (6). The reciprocal is $\frac{6}{1}$. This answer reduces to 6 because $\frac{6}{1} = 6 \div 1 = 6$. Our final answer is 6.

Example: Find the reciprocal of 12.

First rewrite the whole number 12 as a fraction by dividing by one: $12 = \frac{12}{1}$. Now swap the numerator (12) with the denominator (1). The reciprocal is $\frac{1}{12}$. Our final answer is $\frac{1}{12}$.

Example: Find the reciprocal of $\frac{5}{4}$.

Swap the numerator (5) with the denominator (4). The reciprocal is $\frac{4}{5}$.

Example: Find the reciprocal of 50.

First rewrite the whole number 50 as a fraction by dividing by one: $50 = \frac{50}{1}$. Now swap the numerator (50) with the denominator (1). The reciprocal is $\frac{1}{50}$. Our final answer is $\frac{1}{50}$.

Date: _____   Score: ___/8   Name: _____

Chapter 8 Exercises – Part A

Directions: Find the reciprocal of each fraction.

❶

$\dfrac{2}{3}$

❷

$\dfrac{6}{5}$

❸

$\dfrac{1}{3}$

❹

9

❺

$\dfrac{5}{8}$

❻

2

❼

$\dfrac{7}{10}$

❽

$\dfrac{1}{25}$

❖ Check your answers using the Answer Key at the back of the book.

Date: _____   Score: ___/8   Name: _____

Chapter 8 Exercises – Part B

Directions: Find the reciprocal of each fraction.

❾

$$\frac{9}{2}$$

❿

$$5$$

⓫

$$\frac{1}{200}$$

⓬

$$\frac{4}{15}$$

⓭

$$\frac{1}{4}$$

⓮

$$\frac{3}{4}$$

⓯

$$\frac{9}{16}$$

⓰

$$1$$

❖ Check your answers using the Answer Key at the back of the book.

9 DIVIDING FRACTIONS

Dividing one fraction by another, such as $\frac{1}{2} \div \frac{3}{7}$, is equivalent to multiplying the first fraction $\left(\frac{1}{2}\right)$ by the reciprocal of the second fraction (the reciprocal of $\frac{3}{7}$ is $\frac{7}{3}$). For example, the division problem $\frac{1}{2} \div \frac{3}{7}$ is equivalent to the multiplication problem $\frac{1}{2} \times \frac{7}{3}$, where the second fraction of division $\left(\frac{3}{7}\right)$ was replaced by its reciprocal $\left(\frac{7}{3}\right)$ for the multiplication.

The strategy for dividing two fractions is:
- First find the reciprocal of the second fraction as described in Chapter 8.
- Multiply the first fraction by the reciprocal of the second fraction using the method from Chapter 7: Multiply the two numerators together to form the new numerator and multiply the two denominators together to form the new denominator.
- If the answer can be reduced, apply the method from Chapter 2 in order to reduce the fraction.

For example, consider the two fractions divided below.
$$\frac{1}{2} \div \frac{3}{7} = \frac{1}{2} \times \frac{7}{3} = \frac{1 \times 7}{2 \times 3} = \frac{7}{6}$$

You can find additional examples, along with step-by-step explanations, beginning on the next page.

Definitions

- **numerator**: the number that appears in the top of a fraction.
- **denominator**: the number that appears in the bottom of a fraction.
- **reciprocal**: what you get when you swap the numerator and denominator.

Example: Perform the division $\frac{4}{3} \div \frac{2}{5}$.

First find the reciprocal of the second fraction: The reciprocal of $\frac{2}{5}$ is $\frac{5}{2}$. Note that dividing $\frac{4}{3}$ by $\frac{2}{5}$ is equivalent to multiplying $\frac{4}{3}$ by $\frac{5}{2}$.

$$\frac{4}{3} \div \frac{2}{5} = \frac{4}{3} \times \frac{5}{2}$$

Multiply $\frac{4}{3}$ by $\frac{5}{2}$ following these steps:

- Multiply the numerators (4 and 5) together to form the new numerator.
- Multiply the denominators (3 and 2) together to form the new denominator.

$$\frac{4}{3} \times \frac{5}{2} = \frac{4 \times 5}{3 \times 2} = \frac{20}{6}$$

- The answer can be reduced since 20 and 6 are both divisible by 2. Divide the numerator and denominator both by 2.

$$\frac{20}{6} = \frac{20 \div 2}{6 \div 2} = \frac{10}{3}$$

Our final answer is $\frac{10}{3}$.

Example: Perform the division $\frac{1}{8} \div \frac{5}{7}$.

First find the reciprocal of the second fraction: The reciprocal of $\frac{5}{7}$ is $\frac{7}{5}$. Note that dividing $\frac{1}{8}$ by $\frac{5}{7}$ is equivalent to multiplying $\frac{1}{8}$ by $\frac{7}{5}$.

$$\frac{1}{8} \div \frac{5}{7} = \frac{1}{8} \times \frac{7}{5}$$

Multiply $\frac{1}{8}$ by $\frac{7}{5}$ following these steps:

- Multiply the numerators (1 and 7) together to form the new numerator.
- Multiply the denominators (8 and 5) together to form the new denominator.

$$\frac{1}{8} \times \frac{7}{5} = \frac{1 \times 7}{8 \times 5} = \frac{7}{40}$$

Our final answer is $\frac{7}{40}$.

Example: Perform the division $\frac{8}{9} \div \frac{2}{3}$.

First find the reciprocal of the second fraction: The reciprocal of $\frac{2}{3}$ is $\frac{3}{2}$. Note that dividing $\frac{8}{9}$ by $\frac{2}{3}$ is equivalent to multiplying $\frac{8}{9}$ by $\frac{3}{2}$.

$$\frac{8}{9} \div \frac{2}{3} = \frac{8}{9} \times \frac{3}{2}$$

Multiply $\frac{8}{9}$ by $\frac{3}{2}$ following these steps:

- Multiply the numerators (8 and 3) together to form the new numerator.
- Multiply the denominators (9 and 2) together to form the new denominator.

$$\frac{8}{9} \times \frac{3}{2} = \frac{8 \times 3}{9 \times 2} = \frac{24}{18}$$

- The answer can be reduced since 24 and 18 are both divisible by 6. Divide the numerator and denominator both by 6.

$$\frac{24}{18} = \frac{24 \div 6}{18 \div 6} = \frac{4}{3}$$

Our final answer is $\frac{4}{3}$.

Example: Perform the division $\frac{15}{4} \div \frac{7}{3}$.

First find the reciprocal of the second fraction: The reciprocal of $\frac{7}{3}$ is $\frac{3}{7}$. Note that dividing $\frac{15}{4}$ by $\frac{7}{3}$ is equivalent to multiplying $\frac{15}{4}$ by $\frac{3}{7}$.

$$\frac{15}{4} \div \frac{7}{3} = \frac{15}{4} \times \frac{3}{7}$$

Multiply $\frac{15}{4}$ by $\frac{3}{7}$ following these steps:

- Multiply the numerators (15 and 3) together to form the new numerator.
- Multiply the denominators (4 and 7) together to form the new denominator.

$$\frac{15}{4} \times \frac{3}{7} = \frac{15 \times 3}{4 \times 7} = \frac{45}{28}$$

Our final answer is $\frac{45}{28}$. (Note that 45 and 28 aren't evenly divisible by any common factors.)

You can use the same method to perform division between a whole number and a fraction. For example, consider dividing 5 by $\frac{5}{4}$. We can write the whole number over a denominator of one, applying the rule that $\frac{5}{1} = 5$. We learned this in Chapter 2 (see page 20): We can think of $\frac{5}{1}$ as being $5 \div 1$, which is equal to 5.

Example: Perform the division $5 \div \frac{5}{4}$.

First rewrite the whole number 5 as a fraction by dividing by one: $5 = \frac{5}{1}$. Now we can express the division problem using two fractions: $\frac{5}{1} \div \frac{5}{4}$.

Next find the reciprocal of the second fraction: The reciprocal of $\frac{5}{4}$ is $\frac{4}{5}$. Note that dividing $\frac{5}{1}$ by $\frac{5}{4}$ is equivalent to multiplying $\frac{5}{1}$ by $\frac{4}{5}$.

$$\frac{5}{1} \div \frac{5}{4} = \frac{5}{1} \times \frac{4}{5}$$

Multiply $\frac{5}{1}$ by $\frac{4}{5}$ following these steps:

- Multiply the numerators (5 and 4) together to form the new numerator.
- Multiply the denominators (1 and 5) together to form the new denominator.

$$\frac{5}{1} \times \frac{4}{5} = \frac{5 \times 4}{1 \times 5} = \frac{20}{5}$$

- The answer can be reduced since 20 and 5 are both divisible by 5. Divide the numerator and denominator both by 5.

$$\frac{20}{5} = \frac{20 \div 5}{5 \div 5} = \frac{4}{1} = 4$$

Our final answer is 4. Note that $\frac{4}{1} = 4 \div 1 = 4$. (Alternatively, go back a step and note that $\frac{20}{5} = 20 \div 5 = 4$.)

This solution is shorter if you notice that the 5's cancel (since $\frac{5}{5} = 5 \div 5 = 1$) in $\frac{5}{1} \times \frac{4}{5}$, such that $\frac{5}{1} \div \frac{5}{4} = \frac{5}{1} \times \frac{4}{5} = \frac{4}{1} = 4$.

Example: Perform the division $\frac{9}{8} \div 3$.

First rewrite the whole number 3 as a fraction by dividing by one: $3 = \frac{3}{1}$. Now we can express the division problem using two fractions: $\frac{9}{8} \div \frac{3}{1}$.

Next find the reciprocal of the second fraction: The reciprocal of $\frac{3}{1}$ is $\frac{1}{3}$. Note that dividing $\frac{9}{8}$ by $\frac{3}{1}$ is equivalent to multiplying $\frac{9}{8}$ by $\frac{1}{3}$.

$$\frac{9}{8} \div \frac{3}{1} = \frac{9}{8} \times \frac{1}{3}$$

Multiply $\frac{9}{8}$ by $\frac{1}{3}$ following these steps:

- Multiply the numerators (9 and 1) together to form the new numerator.
- Multiply the denominators (8 and 3) together to form the new denominator.

$$\frac{9}{8} \times \frac{1}{3} = \frac{9 \times 1}{8 \times 3} = \frac{9}{24}$$

- The answer can be reduced since 9 and 24 are both divisible by 3. Divide the numerator and denominator both by 3.

$$\frac{9}{24} = \frac{9 \div 3}{24 \div 3} = \frac{3}{8}$$

Our final answer is $\frac{3}{8}$.

Date: _____   Score: ___/5   Name: _____

Chapter 9 Exercises – Part A

Directions: Divide the given fractions. Express your answer as a reduced fraction.

❶

$$\frac{6}{7} \div \frac{5}{2} =$$

❷

$$\frac{3}{4} \div \frac{1}{2} =$$

❸

$$\frac{6}{5} \div \frac{3}{7} =$$

❹

$$6 \div \frac{5}{3} =$$

❺

$$\frac{1}{2} \div \frac{1}{4} =$$

❖ Check your answers using the Answer Key at the back of the book.

Date: _____   Score: ___/5   Name: _____

Chapter 9 Exercises – Part B

Directions: Divide the given fractions. Express your answer as a reduced fraction.

❻

$$\frac{3}{8} \div \frac{4}{3} =$$

❼

$$\frac{4}{3} \div \frac{20}{9} =$$

❽

$$\frac{16}{7} \div \frac{8}{3} =$$

❾

$$\frac{3}{4} \div 2 =$$

❿

$$\frac{5}{6} \div \frac{1}{6} =$$

❖ Check your answers using the Answer Key at the back of the book.

Date: _____   Score: ___/5   Name: _____

Chapter 9 Exercises – Part C

Directions: Divide the given fractions. Express your answer as a reduced fraction.

⓫

$$\frac{3}{16} \div \frac{9}{4} =$$

⓬

$$\frac{2}{5} \div \frac{5}{2} =$$

⓭

$$\frac{6}{5} \div \frac{18}{25} =$$

⓮

$$\frac{5}{8} \div \frac{5}{8} =$$

⓯

$$8 \div \frac{2}{3} =$$

❖ Check your answers using the Answer Key at the back of the book.

10 DECIMALS

A decimal is an alternative way of expressing a fraction. A decimal is basically a fraction that has a denominator of 10, 100, 1000, or another power of 10. For example, 0.7 means $\frac{7}{10}$ and 0.43 means $\frac{43}{100}$.

The denominator of the decimal is determined by the place value. The table below indicates the place value of each digit for a numerical example.

$$1\,,2\quad 3\quad 4\,.\,5\quad 6\quad 7$$

| thousands | hundreds | tens | units | tenths | hundredths | thousandths |

The number shown above (which is 1,234.567) has three decimal places. The digits which follow the decimal point (.567) are the decimal places.

- The 5 is in the "tenths" place.
- The 6 is in the "hundredths" place.
- The 7 is in the "thousandths" place.

Place values to the right of the decimal point end with "ths." For example, compare the "tenths" place to the "tens" place, or the "hundredths" place to the "hundreds" place. The letters "th" near the end make a very significant difference.

Definition

- **decimal**: an alternative way to express a fraction, where the denominator is a power of 10. For example, 0.2 means $\frac{2}{10}$ and 0.37 means $\frac{37}{100}$.

A decimal can be converted into a fraction by following these steps:
- First remove the decimal point. If the number begins with zeros, as in 0.04 or 0.97, ignore the leading zeros. For example, 0.04 would change into 4 and 0.97 would change into 97.
- Next divide by a power of 10 corresponding to the final decimal position. Consult the table on the previous page to determine the final decimal place. For example, the final digit of 16.473 is a 3 and appears in the thousandths place, so we would divide by 1000. As another example, the final digit of 0.8 is an 8 and appears in the tenths place, so we would divide by 10.
- If the answer can be reduced, apply the method from Chapter 2 in order to reduce the fraction.

This strategy is illustrated in the examples that follow.

Example: Convert the decimal 0.25 into a fraction.
- Remove the decimal point from 0.25 to get 25. (Ignore the leading zero.)
- The final digit (5) of 0.25 is in the hundredths place. Therefore, we divide 25 by one hundred to get $\frac{25}{100}$.
- The answer can be reduced since 25 and 100 are both divisible by 25. Divide the numerator and denominator both by 25.

$$\frac{25}{100} = \frac{25 \div 25}{100 \div 25} = \frac{1}{4}$$

The decimal 0.25 is equivalent to the fraction $\frac{1}{4}$.

Example: Convert the decimal 1.7 into a fraction.
- Remove the decimal point from 1.7 to get 17. (Since 1.7 doesn't begin with a 0, there is nothing to ignore.)
- The final digit (7) of 1.7 is in the tenths place. Therefore, we divide 17 by ten to get $\frac{17}{10}$. This answer can't be reduced since 17 and 10 don't share any common factors. Therefore, $\frac{17}{10}$ is our final answer.

The decimal 1.7 is equivalent to the fraction $\frac{17}{10}$.

(Note: In the United States, we place a point between the units place and the decimal places. In some countries, it is more common to use a comma instead.)

Example: Convert the decimal 0.005 into a fraction.
- Remove the decimal point from 0.005 to get 5. (Ignore the leading zeros.)
- The final digit (5) of 0.005 is in the thousandths place. Therefore, we divide 5 by one thousand to get $\frac{5}{1000}$.
- The answer can be reduced since 5 and 1000 are both divisible by 5. Divide the numerator and denominator both by 5.

$$\frac{5}{1000} = \frac{5 \div 5}{1000 \div 5} = \frac{1}{200}$$

The decimal 0.005 is equivalent to the fraction $\frac{1}{200}$.

Example: Convert the decimal 1.44 into a fraction.
- Remove the decimal point from 1.44 to get 144. (Since 1.44 doesn't begin with a 0, there is nothing to ignore.)
- The final digit (4) of 1.44 is in the hundredths place. Therefore, we divide 144 by one hundred to get $\frac{144}{100}$.
- The answer can be reduced since 144 and 100 are both divisible by 4. Divide the numerator and denominator both by 4.

$$\frac{144}{100} = \frac{144 \div 4}{100 \div 4} = \frac{36}{25}$$

The decimal 1.44 is equivalent to the fraction $\frac{36}{25}$.

Example: Convert the decimal 0.3 into a fraction.
- Remove the decimal point from 0.3 to get 3. (Ignore the leading zero.)
- The final digit (3) of 0.3 is in the tenths place. Therefore, we divide 3 by ten to get $\frac{3}{10}$. This answer can't be reduced since 3 and 10 don't share any common factors. Therefore, $\frac{3}{10}$ is our final answer.

The decimal 0.3 is equivalent to the fraction $\frac{3}{10}$.

One way to convert a fraction into a decimal is to make a common denominator of 10, 100, 1000, or other power of 10, using the technique from Chapter 3. Recall that you may multiply both the numerator and denominator of the fraction by the same value to obtain an equivalent fraction.

For example, consider the fraction $\frac{9}{20}$. We can make an equivalent fraction with a denominator of 100 by multiplying both the numerator (9) and denominator (20) by 5 (since $5 \times 20 = 100$).

$$\frac{9}{20} = \frac{9 \times 5}{20 \times 5} = \frac{45}{100}$$

Once the denominator is a power of 10, we can express the fraction as a decimal. Write the numerator as a decimal such that its final digit ends in the decimal position corresponding to the denominator. For the fraction $\frac{45}{100}$, the final digit of the numerator (which is a 5) must fall in the hundredths place because the denominator is 100. The corresponding decimal is 0.45 (forty-five hundredths).

The steps of this strategy are outlined below.

- What is the smallest power of 10 that is evenly divisible by the denominator? As examples, a denominator of 5 can be changed into 10 since $5 \times 2 = 10$, a denominator of 25 can be changed into 100 since $25 \times 4 = 100$, and a denominator of 200 can be changed into 1000 since $200 \times 5 = 1000$.
- Multiply both the numerator and denominator by the same number in order to make an equivalent fraction with the desired new denominator (the answer to the question in the previous step). The new denominator must be a power of 10 (such as 10, 100, 1000, or 10,000).
- Write this equivalent fraction as a decimal. The final digit of the numerator must match the decimal position corresponding to the denominator. As examples, $\frac{2}{10}$ equals 0.2 (the 2 falls in the tenths place), $\frac{49}{100}$ equals 0.49 (the 9 falls in the hundredths place), and $\frac{711}{1000}$ equals 0.711 (the final 1 falls in the thousandths place).

This strategy is illustrated in the examples that follow.

Example: Convert the fraction $\frac{4}{5}$ into a decimal.

- What is the smallest power of 10 that is evenly divisible by the denominator? The denominator (5) evenly divides into 10 since $5 \times 2 = 10$.

- Multiply both the numerator and denominator of $\frac{4}{5}$ by 2 in order to make an equivalent fraction with a denominator of 10.

$$\frac{4}{5} = \frac{4 \times 2}{5 \times 2} = \frac{8}{10}$$

- Write the new fraction $\frac{8}{10}$ as a decimal with the final digit of the new numerator (8) in the tenths place (corresponding to $\frac{1}{10}$).

The fraction $\frac{4}{5}$ is equivalent to the decimal 0.8 (since $0.8 = \frac{8}{10} = \frac{4}{5}$).

Example: Convert the fraction $\frac{8}{25}$ into a decimal.

- What is the smallest power of 10 that is evenly divisible by the denominator? The denominator (25) evenly divides into 100 since $25 \times 4 = 100$.

- Multiply both the numerator and denominator of $\frac{8}{25}$ by 4 in order to make an equivalent fraction with a denominator of 100.

$$\frac{8}{25} = \frac{8 \times 4}{25 \times 4} = \frac{32}{100}$$

- Write the new fraction $\frac{32}{100}$ as a decimal with the final digit of the new numerator (32) in the hundredths place (corresponding to $\frac{1}{100}$).

The fraction $\frac{8}{25}$ is equivalent to the decimal 0.32 (since $0.32 = \frac{32}{100} = \frac{8}{25}$).

Example: Convert the fraction $\frac{5347}{1000}$ into a decimal.

- What is the smallest power of 10 that is evenly divisible by the denominator? The denominator (1000) is already a power of 10, so we don't need to change it. In this case, we may skip the first two steps.

- Write the fraction $\frac{5347}{1000}$ as a decimal with the final digit of the numerator (5,347) in the thousandths place (corresponding to $\frac{1}{1000}$).

The fraction $\frac{5347}{1000}$ is equivalent to the decimal 5.347 (that is, $5.347 = \frac{5347}{1000}$).

Example: Convert the fraction $\frac{9}{4}$ into a decimal.

- What is the smallest power of 10 that is evenly divisible by the denominator? The denominator (4) evenly divides into 100 since $4 \times 25 = 100$. (Note that 4 doesn't evenly divide into 10.)

- Multiply both the numerator and denominator of $\frac{9}{4}$ by 25 in order to make an equivalent fraction with a denominator of 100.

$$\frac{9}{4} = \frac{9 \times 25}{4 \times 25} = \frac{225}{100}$$

- Write the new fraction $\frac{225}{100}$ as a decimal with the final digit of the new numerator (225) in the hundredths place (corresponding to $\frac{1}{100}$).

The fraction $\frac{9}{4}$ is equivalent to the decimal 2.25 (since $2.25 = \frac{225}{100} = \frac{9}{4}$).

Example: Convert the fraction $\frac{123}{500}$ into a decimal.

- What is the smallest power of 10 that is evenly divisible by the denominator? The denominator (500) evenly divides into 1000 since $500 \times 2 = 1000$.

- Multiply both the numerator and denominator of $\frac{123}{500}$ by 2 in order to make an equivalent fraction with a denominator of 1000.

$$\frac{123}{500} = \frac{123 \times 2}{500 \times 2} = \frac{246}{1000}$$

- Write the new fraction $\frac{246}{1000}$ as a decimal with the final digit of the new numerator (246) in the thousandths place (corresponding to $\frac{1}{1000}$).

The fraction $\frac{123}{500}$ is equivalent to the decimal 0.246 (since $0.246 = \frac{246}{1000} = \frac{123}{500}$).

Notes:

- A power of 10 in the denominator effectively moves the decimal point. For example, to write $\frac{2481}{1000}$ as a decimal, you can move the decimal point 3 places to the left (one for each zero in 1000) to get 2.481 from 2481.

- An alternative method for converting a fraction to a decimal involves the method of long division. The method we're using in this chapter is simpler. (If you're interested in how long division can be used to turn fractions into decimals, see Chapter 20.)

Date: _____   Score: ___/5   Name: _____

Chapter 10 Exercises – Part A

Directions: Convert each decimal into a reduced fraction.

❶

　　0.5

❷

　　0.75

❸

　　1.1

❹

　　0.325

❺

　　1.75

❖ Check your answers using the Answer Key at the back of the book.

Date: _____   Score: ___/5   Name: _____

Chapter 10 Exercises – Part B

Directions: Convert each decimal into a reduced fraction.

❻

0.9

❼

0.05

❽

2.5

❾

0.002

❿

1.2

❖ Check your answers using the Answer Key at the back of the book.

Date: _____   Score: ___/5   Name: _____

Chapter 10 Exercises – Part C

Directions: Convert each fraction into a decimal.

⓫

$$\frac{3}{2}$$

⓬

$$\frac{3}{5}$$

⓭

$$\frac{11}{20}$$

⓮

$$\frac{5}{4}$$

⓯

$$\frac{7}{250}$$

❖ Check your answers using the Answer Key at the back of the book.

Date: _____   Score: ___/5   Name: _____

Chapter 10 Exercises – Part D

Directions: Convert each fraction into a decimal.

⓰

$$\frac{71}{50}$$

⓱

$$\frac{17}{10}$$

⓲

$$\frac{16}{25}$$

⓳

$$\frac{3}{8}$$

⓴

$$\frac{9}{200}$$

❖ Check your answers using the Answer Key at the back of the book.

11 REPEATING DECIMALS

When some fractions are converted to decimals, they result in repeated digits. For example, the fraction $\frac{1}{3}$ equals 0.333333... with the 3's repeating forever.

(Why? It has to do with long division. If you multiply $3 \times 0.3 = 0.9$, it's 0.1 less than 1. Then if you multiply $3 \times 0.03 = 0.09$, it's 0.01 less than 0.1. Next if you multiply $3 \times 0.003 = 0.009$, it's 0.001 less than 0.01. This continues forever, as shown below.)

$$
\begin{array}{r}
0.333 \\
3{\overline{)\,1.0}} \\
-0.9 \\
\hline
0.10 \\
-0.09 \\
\hline
0.010 \\
-0.009 \\
\hline
0.001
\end{array}
$$

Note: If you would like to better understand the long division example above, read Chapter 20. Another way to see that $\frac{1}{3} = 0.333333...$ is to enter $1 \div 3$ on your calculator.

Definitions

- **decimal**: an alternative way to express a fraction, where the denominator is a power of 10. For example, 0.2 means $\frac{2}{10}$ and 0.37 means $\frac{37}{100}$.
- **repeating decimal**: a decimal that has repeated values. A bar is placed over the digits that repeat. For example, $0.\overline{7}$ means 0.777777... and $0.\overline{45}$ means 0.454545... with the 7's and 45's repeating forever.

Let's focus on identifying and working with repeating decimals. The important point is that we place a bar over the digits that repeat. Consider the following examples.

- $0.\overline{2}$ means 0.222222... with the 2's repeating forever.
- $0.\overline{37}$ means 0.373737... with the 37's repeating forever.
- $0.\overline{845}$ means 0.845845... with the 845's repeating forever.
- $1.\overline{4}$ means 1.444444... with the 4's repeating forever.
- $0.5\overline{8}$ means 0.5888888... with the 8's repeating forever.

It helps to look closely. Compare the two similar, but different, examples below.

- $0.7\overline{1}$ means 0.7111111... with only the 1's repeating forever, but just one 7.
- $0.\overline{71}$ means 0.717171... with the 71's repeating forever.

The numbers $0.7\overline{1}$ and $0.\overline{71}$ look very similar, but the difference is significant. Pay close attention to which number or numbers are under the bar.

A repeating decimal can be converted into a fraction by following these steps:

- Does the repeating decimal contain any digits that don't repeat? For example, the 3 in $0.3\overline{5}$ doesn't repeat (only the 5 does), whereas the 3 and 5 both repeat in $0.\overline{35}$. If so, follow the strategy described on page 91.
- First remove the decimal point. If the number begins with zeros, as in $0.0\overline{4}$ or $0.\overline{97}$, ignore the leading zeros. Also remove the bar over the digits. For example, $0.0\overline{4}$ would change into 4 and $0.\overline{97}$ would change into 97.
- How many digits have a bar over them? If just one digit has a bar over it (like $0.\overline{7}$ or $0.0\overline{4}$), divide the previous number by 9. If two digits have a bar over them (like $0.\overline{32}$ or $0.0\overline{51}$), divide the previous number by 99. If three digits have a bar over them (like $0.\overline{981}$), divide the previous number by 999. Etc.
- How many 0's were between the decimal point and first nonzero digit of the original value? If there were none (like $0.\overline{63}$), skip to the next step. If there was one zero (like $0.0\overline{2}$), divide your previous answer by 10. If there were two zeros (like $0.00\overline{41}$), divide your previous answer by 100. If there were three zeros (like $0.000\overline{7}$), divide your previous answer by 1000. Etc.
- If the answer can be reduced, apply the method from Chapter 2 in order to reduce the fraction.

This strategy is illustrated in the examples that follow.

Example: Convert the decimal $0.\overline{4}$ into a fraction.

- Remove the decimal point from $0.\overline{4}$ to get 4. (Ignore the leading zero and also remove the bar.)
- How many digits have a bar over them in the original value? Just one digit (the 4) has a bar over it in $0.\overline{4}$.
- Since one digit has a bar over it, we divide the answer to the first step (which was 4) by the number 9. This makes the fraction $\frac{4}{9}$.
- There are no 0's between the decimal point and the 4 in $0.\overline{4}$, so we don't need to divide our answer by a power of 10 in this example.
- This answer can't be reduced since 4 and 9 don't share any common factors. Therefore, $\frac{4}{9}$ is our final answer.

The repeating decimal $0.\overline{4}$ is equivalent to the fraction $\frac{4}{9}$.

Check: If you enter $4 \div 9$ on your calculator, you will get 0.44444444...

Example: Convert the decimal $0.\overline{27}$ into a fraction.

- Remove the decimal point from $0.\overline{27}$ to get 27. (Ignore the leading zero and also remove the bar.)
- How many digits have a bar over them in the original value? Two digits (the 2 and 7) have a bar over them in $0.\overline{27}$.
- Since two digits have a bar over them, we divide the answer to the first step (which was 27) by the number 99. This makes the fraction $\frac{27}{99}$.
- There are no 0's between the decimal point and the 2 in $0.\overline{27}$, so we don't need to divide our answer by a power of 10 in this example.
- The answer can be reduced since 27 and 99 are both divisible by 9. Divide the numerator and denominator both by 9.

$$\frac{27}{99} = \frac{27 \div 9}{99 \div 9} = \frac{3}{11}$$

The repeating decimal $0.\overline{27}$ is equivalent to the fraction $\frac{3}{11}$.

Check: If you enter $3 \div 11$ on your calculator, you will get 0.27272727... Note that a calculator may round the last 2 up to a 3 (if the final digit on the display would otherwise have been a 2).

Example: Convert the decimal $0.00\bar{8}$ into a fraction.

- Remove the decimal point from $0.00\bar{8}$ to get 8. (Ignore the leading zeros and also remove the bar.)
- How many digits have a bar over them in the original value? Just one digit (the 8) has a bar over it in $0.00\bar{8}$.
- Since one digit has a bar over it, we divide the answer to the first step (which was 8) by the number 9. This makes the fraction $\frac{8}{9}$.
- There are two 0's between the decimal point and the 8 in $0.00\bar{8}$, so we divide by 100 in this example. It may help to review Chapter 9.

$$\frac{8}{9} \div 100 = \frac{8}{9} \div \frac{100}{1} = \frac{8}{9} \times \frac{1}{100} = \frac{8 \times 1}{9 \times 100} = \frac{8}{900}$$

- The answer can be reduced since 8 and 900 are both divisible by 4. Divide the numerator and denominator both by 4. Note that $900 \div 4 = 225$.

$$\frac{8}{900} = \frac{8 \div 4}{900 \div 4} = \frac{2}{225}$$

The repeating decimal $0.00\bar{8}$ is equivalent to the fraction $\frac{2}{225}$.

Check: If you enter $2 \div 225$ on your calculator, you will get 0.00888888...

Example: Convert the decimal $0.0\overline{523}$ into a fraction.

- Remove the decimal point from $0.0\overline{523}$ to get 523. (Ignore the leading zeros and also remove the bar.)
- How many digits have a bar over them in the original value? Three digits (the 5, 2, and 3) have a bar over them in $0.0\overline{523}$.
- Since three digits have a bar over them, we divide the answer to the first step (which was 523) by the number 999. This makes the fraction $\frac{523}{999}$.
- There is one 0 between the decimal point and the 5 in $0.0\overline{523}$, so we divide by 10 in this example. It may help to review Chapter 9.

$$\frac{523}{999} \div 10 = \frac{523}{999} \div \frac{10}{1} = \frac{523}{999} \times \frac{1}{10} = \frac{523 \times 1}{999 \times 10} = \frac{523}{9990}$$

- This answer can't be reduced since 523 and 9990 don't share any common factors. Therefore, $\frac{523}{9990}$ is our final answer.

The repeating decimal $0.0\overline{523}$ is equivalent to the fraction $\frac{523}{9990}$.

Check: If you enter $523 \div 9990$ on your calculator, you will get 0.0523523523...

If the repeating decimal contains any digits that don't repeat (for example, the 3 in $0.3\overline{5}$ doesn't repeat – only the 5 does – whereas the 3 and 5 both repeat in $0.\overline{35}$), follow these steps:

- Express the given value as the sum of nonrepeating and repeating parts, like the examples below. Be careful to preserve the original decimal positions.
  - $0.3\overline{5} = 0.3 + 0.0\overline{5}$ (The 3 is in the tenths place and the repeating 5 begins in the hundredths place.)
  - $1.\overline{23} = 1 + 0.\overline{23}$ (The 1 is in the units place, the 2 is in the tenths place, and the 3 is in the hundredths place.)
  - $0.04\overline{7} = 0.04 + 0.00\overline{7}$ (The 4 is in the hundredths place and the repeating 7 begins in the thousandths place.)
- Convert the nonrepeating part into a fraction following the method of chapter 10. For example, $0.3 = \frac{3}{10}$ and $0.04 = \frac{4}{100} = \frac{4 \div 4}{100 \div 4} = \frac{1}{25}$. If the non-repeating part is a whole number, divide by 1 to make the fraction. For example, $1 = \frac{1}{1}$ and $5 = \frac{5}{1}$.
- Convert the repeating part into a fraction following the method described on page 88 (and illustrated in examples on pages 89-90).
- Add the two fractions (from the two previous steps) together using a common denominator. It may help to review Chapter 5.
- If the answer can be reduced, apply the method from Chapter 2 in order to reduce the fraction.

This strategy is illustrated in the examples that follow.

Example: Convert the decimal $2.\overline{3}$ into a fraction.

- First separate the given value into nonrepeating and repeating parts. The 2 is in the units place and the repeating 3 begins in the tenths place.

$$2.\overline{3} = 2 + 0.\overline{3}$$

- Express the nonrepeating part (which is the number 2) as a fraction. Since 2 is a whole number, divide by one.

$$2 = \frac{2}{1}$$

- Convert the repeating part $(0.\overline{3})$ into a fraction, following the strategy on page 88.
  - Remove the decimal point from $0.\overline{3}$ to get 3. (Ignore the leading zero and also remove the bar.)
  - How many digits have a bar over them in $0.\overline{3}$? Just one digit (the 3) has a bar over it in $0.\overline{3}$.
  - Since one digit has a bar over it, we divide the number 3 (which we found two steps above) by the number 9. This makes the fraction $\frac{3}{9}$.
  - There are no 0's between the decimal point and the 3 in $0.\overline{3}$, so we don't need to divide our answer by a power of 10 in this example.
  - The fraction $\frac{3}{9}$ can be reduced since 3 and 9 are both divisible by 3. Divide the numerator and denominator both by 3.

$$\frac{3}{9} = \frac{3 \div 3}{9 \div 3} = \frac{1}{3}$$

- Now we add the fractions $\frac{2}{1}$ and $\frac{1}{3}$ (which we found in previous steps). Multiply both the numerator and denominator of $\frac{2}{1}$ by 3 in order to make a common denominator of 3.

$$\frac{2}{1} + \frac{1}{3} = \frac{2 \times 3}{1 \times 3} + \frac{1}{3} = \frac{6}{3} + \frac{1}{3} = \frac{6+1}{3} = \frac{7}{3}$$

- This answer can't be reduced since 7 and 3 don't share any common factors. Therefore, $\frac{7}{3}$ is our final answer.

The repeating decimal $2.\overline{3}$ is equivalent to the fraction $\frac{7}{3}$.

Check: If you enter $7 \div 3$ on your calculator, you will get 2.33333333...

Example: Convert the decimal $0.5\overline{63}$ into a fraction.

- First separate the given value into nonrepeating and repeating parts. The 5 is in the tenths place, the 6 is in the hundredths place, and the 3 is in the thousandths place.

$$0.5\overline{63} = 0.5 + 0.0\overline{63}$$

- Express the nonrepeating part (which is 0.5) as a fraction using the method from Chapter 10. Since the 5 is in the tenths place, divide by 10.

$$0.5 = \frac{5}{10}$$

- Convert the repeating part $(0.0\overline{63})$ into a fraction, following the strategy on page 88.
  - Remove the decimal point from $0.0\overline{63}$ to get 63. (Ignore the leading zeros and also remove the bar.)
  - How many digits have a bar over them in $0.0\overline{63}$? Two digits (the 6 and 3) have a bar over them in $0.0\overline{63}$.
  - Since two digits have a bar over them, we divide the number 63 (which we found two steps above) by the number 99. This makes the fraction $\frac{63}{99}$.
  - There is one 0 between the decimal point and the 6 in $0.0\overline{63}$, so we divide by 10 in this example. It may help to review Chapter 9.

$$\frac{63}{99} \div 10 = \frac{63}{99} \div \frac{10}{1} = \frac{63}{99} \times \frac{1}{10} = \frac{63 \times 1}{99 \times 10} = \frac{63}{990}$$

- Now we add the fractions $\frac{5}{10}$ and $\frac{63}{990}$ (which we found in previous steps). Multiply both the numerator and denominator of $\frac{5}{10}$ by 99 in order to make a common denominator of 99 (since $10 \times 99 = 990$).

$$\frac{5}{10} + \frac{63}{990} = \frac{5 \times 99}{10 \times 99} + \frac{63}{990} = \frac{495}{990} + \frac{63}{990} = \frac{495 + 63}{990} = \frac{558}{990}$$

- The answer can be reduced since 558 and 990 are both divisible by 18 (that's the greatest common factor). Divide the numerator and denominator both by 18. Note that $558 \div 18 = 31$ and $990 \div 18 = 55$.

$$\frac{558}{990} = \frac{558 \div 18}{990 \div 18} = \frac{31}{55}$$

The repeating decimal $0.5\overline{63}$ is equivalent to the fraction $\frac{31}{55}$.

Check: If you enter $31 \div 55$ on your calculator, you will get 0.563636363...

Example: Convert the decimal $0.25\overline{08}$ into a fraction.

- First separate the given value into nonrepeating and repeating parts. The 2 is in the tenths place, the 5 is in the hundredths place, the second 0 is in the thousandths place, and the 8 is in the ten thousandths place.

$$0.25\overline{08} = 0.25 + 0.00\overline{08}$$

- Express the nonrepeating part (which is 0.25) as a fraction using the method from Chapter 10. Since the 5 is in the hundredths place, divide by 100.

$$0.25 = \frac{25}{100}$$

- Convert the repeating part ($0.00\overline{08}$) into a fraction, following the strategy on page 88. Note that $0.00\overline{08}$ means $0.000808080808\ldots$ (The 08 repeats.)
  - Remove the decimal point from $0.00\overline{08}$ to get 8. (Ignore the leading zeros and also remove the bar.)
  - How many digits have a bar over them in $0.00\overline{08}$? Two digits (one 0 and the 8) have a bar over them in $0.00\overline{08}$.
  - Since two digits have a bar over them, we divide the number 8 (which we found two steps above) by the number 99. This makes the fraction $\frac{8}{99}$.
  - There are two 0's between the decimal point and the repeating part of $0.00\overline{08}$ (the 08 with the bar is repeating), so we divide by 100 in this example. It may help to review Chapter 9.

$$\frac{8}{99} \div 100 = \frac{8}{99} \div \frac{100}{1} = \frac{8}{99} \times \frac{1}{100} = \frac{8 \times 1}{99 \times 100} = \frac{8}{9900}$$

- Now we add the fractions $\frac{25}{100}$ and $\frac{8}{9900}$ (which we found in previous steps). Multiply both the numerator and denominator of $\frac{25}{100}$ by 99 in order to make a common denominator of 9900 (since $100 \times 99 = 9900$). Note that $25 \times 99 = 2475$.

$$\frac{25}{100} + \frac{8}{9900} = \frac{25 \times 99}{100 \times 99} + \frac{8}{9900} = \frac{2475}{9900} + \frac{8}{9900} = \frac{2483}{9900}$$

- This answer can't be reduced since 2483 and 9900 don't share any common factors (see the tips on page 17). Therefore, $\frac{2483}{9900}$ is our final answer.

The repeating decimal $0.25\overline{08}$ is equivalent to the fraction $\frac{2483}{9900}$.

Check: If you enter $2483 \div 9900$ on your calculator, you will get $0.2508080808\ldots$

A fraction converts into a repeating decimal if the denominator can't be factored exclusively with powers of 2 and 5. (Why 2 and 5? It's because $2 \times 5 = 10$, and since the standard number system is based on the number 10.)

For example, $\frac{1}{40}$ doesn't result in a repeating decimal because the denominator (40) factors as $2 \times 2 \times 2 \times 5$ (which is the same as 8×5). If you use a calculator, you can see that $\frac{1}{40} = 0.025$. As a counterexample, $\frac{1}{30}$ results in a repeating decimal because the denominator (30) factors as $2 \times 3 \times 5$, which has a number besides 2 and 5 (it has a 3). If you use a calculator, you can see that $\frac{1}{30} = 0.033333333...$

If a fraction converts into a repeating decimal, one way to do the conversion is to make a new denominator of 9, 90, 99, 900, 990, 999, etc. using the technique from Chapter 3. Recall that you may multiply both the numerator and denominator of the fraction by the same value to obtain an equivalent fraction.

For example, consider the fraction $\frac{8}{15}$. We can make an equivalent fraction with a denominator of 90 by multiplying both the numerator (8) and denominator (15) by 6 (since $15 \times 6 = 90$).

$$\frac{8}{15} = \frac{8 \times 6}{15 \times 6} = \frac{48}{90}$$

Once the denominator is suitable, the fraction can be expressed as a repeating decimal. The strategy outlined on the following page will show you how.

A fraction can be converted into a repeating decimal by following these steps:

- Note: If the denominator factors exclusively in terms of 2's and 5's, such as $20 = 2 \times 2 \times 5$ or $50 = 2 \times 5 \times 5$, the fraction won't make a repeating decimal. In that case, you should consult Chapter 10 instead.

- Which of the following numbers is the smallest denominator that you can make from the given denominator? 9, 90, 99, 900, 990, 999, 9000, 9900, 9990, 9999, etc. Note that the 0's come at the end (unlike 909). For example, a denominator of 3 can be turned into 9 since $3 \times 3 = 9$, whereas a denominator of 225 can be turned into 900 since $225 \times 4 = 900$.

- Multiply both the numerator and denominator by the same number in order to make an equivalent fraction with the desired new denominator (the answer to the question in the previous step).

- Once you have an equivalent fraction with the desired denominator, write a repeating decimal. The number of 9's in the denominator tells you how many digits of the numerator repeat. For example, $\frac{2}{9} = 0.\overline{2}$ has one digit repeating, $\frac{45}{99} = 0.\overline{45}$ has two digits repeating, and $\frac{857}{999} = 0.\overline{857}$ has three digits repeating. Before you write the repeating decimal, check the following special cases:

  - If the denominator ends with 0, first remove the zeros and solve the problem the same way that you would if the zeros weren't in the denominator. When you finish, shift the decimal point one place to the left for each 0 that you removed. For example, you would rewrite $\frac{7}{90}$ as $\frac{7}{9}$, determine that $\frac{7}{9} = 0.\overline{7}$, and shift the decimal point one place to the left to get $\frac{7}{90} = 0.0\overline{7}$. Similarly, you would rewrite $\frac{56}{9900}$ as $\frac{56}{99}$, determine that $\frac{56}{99} = 0.\overline{56}$, and shift the decimal point two places to the left to get $\frac{56}{9900} = 0.00\overline{56}$.

  - If the numerator has fewer digits than the number of 9's in the denominator, there will be one repeating 0 for each missing digit. For example, $\frac{8}{999} = 0.\overline{008}$ and $\frac{73}{999} = 0.\overline{073}$.

  - If the numerator is larger than the denominator, write the numerator as a whole number plus a proper fraction, as demonstrated in the last two examples. Then write the proper fraction of the mixed number as a repeating decimal, and combine it with the whole number.

Example: Convert the fraction $\frac{1}{3}$ into a repeating decimal.

- From the given denominator (3), we can make a denominator of 9. Multiply both the numerator and denominator of $\frac{1}{3}$ by 3 in order to make an equivalent fraction with a denominator of 9.

$$\frac{1}{3} = \frac{1 \times 3}{3 \times 3} = \frac{3}{9}$$

- None of the special cases apply to $\frac{3}{9}$ since the denominator doesn't have a 0, the numerator has the same number of digits as the number of 9's in the denominator, and the numerator is smaller than the denominator.

- Write the fraction $\frac{3}{9}$ as a decimal with the only digit of the numerator repeating: $\frac{3}{9} = 0.\overline{3}$.

The fraction $\frac{1}{3}$ is equivalent to the repeating decimal $0.\overline{3}$ (since $0.\overline{3} = \frac{3}{9} = \frac{1}{3}$).

Check: If you enter $1 \div 3$ on your calculator, you will get 0.33333333...

Example: Convert the fraction $\frac{2}{45}$ into a repeating decimal.

- From the given denominator (45), we can make a denominator of 90. Multiply both the numerator and denominator of $\frac{2}{45}$ by 2 in order to make an equivalent fraction with a denominator of 90.

$$\frac{2}{45} = \frac{2 \times 2}{45 \times 2} = \frac{4}{90}$$

- The new fraction $\frac{4}{90}$ is one of the special cases: The denominator has a 0. First ignore the 0 in the denominator to get $\frac{4}{9}$. Next write $\frac{4}{9}$ as a decimal with the only digit of the numerator repeating: $\frac{4}{9} = 0.\overline{4}$. Finally, shift the decimal point one place to the left to make up for the 0 that was removed from $\frac{4}{90}$ (when we found that $\frac{4}{9} = 0.\overline{4}$). When we shift the decimal point, we get $\frac{4}{90} = 0.0\overline{4}$.

The fraction $\frac{2}{45}$ is equivalent to the repeating decimal $0.0\overline{4}$ (since $0.0\overline{4} = \frac{4}{90} = \frac{2}{45}$).

Check: If you enter $2 \div 45$ on your calculator, you will get 0.044444444...

Example: Convert the fraction $\frac{7}{11}$ into a repeating decimal.

- From the given denominator (11), we can make a denominator of 99. Multiply both the numerator and denominator of $\frac{7}{11}$ by 9 in order to make an equivalent fraction with a denominator of 99.

$$\frac{7}{11} = \frac{7 \times 9}{11 \times 9} = \frac{63}{99}$$

- None of the special cases apply to $\frac{63}{99}$ since the denominator doesn't have a 0, the numerator has the same number of digits as the number of 9's in the denominator, and the numerator is smaller than the denominator.

- Write the fraction $\frac{63}{99}$ as a decimal with the two digits of the numerator both repeating: $\frac{63}{99} = 0.\overline{63}$.

The fraction $\frac{7}{11}$ is equivalent to the repeating decimal $0.\overline{63}$ (since $0.\overline{63} = \frac{63}{99} = \frac{7}{11}$).

Check: If you enter $7 \div 11$ on your calculator, you will get $0.63636363...$

Example: Convert the fraction $\frac{16}{333}$ into a repeating decimal.

- From the given denominator (333), we can make a denominator of 999. Multiply both the numerator and denominator of $\frac{16}{333}$ by 3 in order to make an equivalent fraction with a denominator of 999.

$$\frac{16}{333} = \frac{16 \times 3}{333 \times 3} = \frac{48}{999}$$

- The new fraction $\frac{48}{999}$ is one of the special cases: The numerator has fewer digits than the number of 9's in the denominator. Since the numerator (48) has 2 digits and the denominator (999) has 3 digits, there will be one repeating zero. Write $\frac{48}{999}$ as a decimal with three repeating digits (one for each 9 in 999), including one repeating zero: $\frac{48}{999} = 0.\overline{048}$.

The fraction $\frac{16}{333}$ is equivalent to the repeating decimal $0.\overline{048}$ (since $0.\overline{048} = \frac{48}{999} = \frac{16}{333}$). In the number $0.\overline{048}$, all three digits (the 0, 4, and 8) under the bar repeat.

Check: If you enter $16 \div 333$ on your calculator, you will get $0.048048048048...$

Example: Convert the fraction $\frac{842}{999}$ into a repeating decimal.

- The given denominator (999) happens to already be one of the denominators that we wish to make. We don't need to make an equivalent fraction.
- None of the special cases apply to $\frac{842}{999}$ since the denominator doesn't have a 0, the numerator has the same number of digits as the number of 9's in the denominator, and the numerator is smaller than the denominator.
- Write the fraction $\frac{842}{999}$ as a decimal with all three digits of the numerator repeating: $\frac{842}{999} = 0.\overline{842}$.

The fraction $\frac{842}{999}$ is equivalent to the repeating decimal $0.\overline{842}$.

Check: If you enter 842 ÷ 999 on your calculator, you will get 0.842842842...

Example: Convert the fraction $\frac{7}{3}$ into a repeating decimal.

- From the given denominator (3), we can make a denominator of 9. Multiply both the numerator and denominator of $\frac{7}{3}$ by 3 in order to make an equivalent fraction with a denominator of 9.

$$\frac{7}{3} = \frac{7 \times 3}{3 \times 3} = \frac{21}{9}$$

- The new fraction $\frac{21}{9}$ is one of the special cases: The numerator (21) is larger than the denominator (9). We need to write $\frac{21}{9}$ as a whole number plus a fraction that is smaller than one.
- The way to do this is to divide 21 ÷ 9. The answer is 2 with a remainder of 3 (since $9 \times 2 = 18$ and $21 - 18 = 3$). Therefore, $\frac{21}{9}$ equals 2 plus $\frac{3}{9}$ (the whole number plus the remainder divided by 9): $\frac{21}{9} = 2 + \frac{3}{9}$. If you would like more help with this step, read Chapter 16.
- Write the fraction $\frac{3}{9}$ as a decimal with the only digit of the numerator repeating: $\frac{3}{9} = 0.\overline{3}$.
- Add the repeating decimal $(0.\overline{3})$ from the previous step to the whole number (2) from two steps above to get $2 + 0.\overline{3} = 2.\overline{3}$. This is our final answer.

The fraction $\frac{7}{3}$ is equivalent to the repeating decimal $2.\overline{3}$ (since $2.\overline{3} = 2 + \frac{3}{9} = \frac{21}{9} = \frac{7}{3}$).

Check: If you enter 7 ÷ 3 on your calculator, you will get 2.33333333...

Example: Convert the fraction $\frac{1}{60}$ into a repeating decimal.

- From the given denominator (60), we can make a denominator of 900. Multiply both the numerator and denominator of $\frac{1}{60}$ by 15 in order to make an equivalent fraction with a denominator of 900. Note that $60 \times 15 = 900$.

$$\frac{1}{60} = \frac{1 \times 15}{60 \times 15} = \frac{15}{900}$$

- The new fraction $\frac{15}{900}$ is one of the special cases: The denominator has 0's. First ignore the 0's in the denominator to get $\frac{15}{9}$. We will deal with the 0's that we're removing in a later step. For now, focus on converting $\frac{15}{9}$ into a decimal.

- The fraction $\frac{15}{9}$ is another one of the special cases: The numerator (15) is larger than the denominator (9). We need to write $\frac{15}{9}$ as a whole number plus a fraction that is smaller than one.

- The way to do this is to divide $15 \div 9$. The answer is 1 with a remainder of 6 (since $9 \times 1 = 9$ and $15 - 9 = 6$). Therefore, $\frac{15}{9}$ equals 1 plus $\frac{6}{9}$ (the whole number plus the remainder divided by 9): $\frac{15}{9} = 1 + \frac{6}{9}$. If you would like more help with this step, read Chapter 16.

- Write the fraction $\frac{6}{9}$ as a decimal with the only digit of the numerator repeating: $\frac{6}{9} = 0.\overline{6}$.

- Add the repeating decimal ($0.\overline{6}$) from the previous step to the whole number (1) from two steps above to get $1 + 0.\overline{6} = 1.\overline{6}$. We have found that $\frac{15}{9} = 1.\overline{6}$.

- Finally, shift the decimal point two places to the left to make up for the two 0's that were removed from $\frac{15}{900}$ (when we found that $\frac{15}{9} = 1.\overline{6}$). When we shift the decimal point, we get $\frac{15}{900} = 0.01\overline{6}$. This is our final answer.

The fraction $\frac{1}{60}$ is equivalent to the repeating decimal $0.01\overline{6}$.

Check: If you enter $1 \div 60$ on your calculator, you will get 0.0166666666... Note that a calculator will round the last 6 up to a 7.

Date: _____   Score: ___/5   Name: _____

Chapter 11 Exercises – Part A

Directions: Convert each repeating decimal into a reduced fraction.

❶

$0.\overline{6}$

❷

$0.\overline{63}$

❸

$0.\overline{32}$

❹

$0.0\overline{8}$

❺

$0.\overline{592}$

❖ Check your answers using the Answer Key at the back of the book.

Date: _____   Score: ___/5   Name: _____

Chapter 11 Exercises – Part B

Directions: Convert each repeating decimal into a reduced fraction.

❻

$1.\overline{3}$

❼

$0.01\overline{6}$

❽

$0.2\overline{3}$

❾

$0.\overline{025}$

❿

$0.4\overline{85}$

❖ Check your answers using the Answer Key at the back of the book.

Date: _____   Score: ___/5   Name: _____

Chapter 11 Exercises – Part C

Directions: Convert each fraction into a repeating decimal.

⑪

$$\frac{2}{3}$$

⑫

$$\frac{1}{300}$$

⑬

$$\frac{100}{111}$$

⑭

$$\frac{8}{3}$$

⑮

$$\frac{2}{333}$$

❖ Check your answers using the Answer Key at the back of the book.

Date: _____ Score: ___/5 Name: _____

Chapter 11 Exercises – Part D

Directions: Convert each fraction into a repeating decimal.

16

$$\frac{10}{9}$$

17

$$\frac{5}{33}$$

18

$$\frac{4}{45}$$

19

$$\frac{8}{15}$$

20

$$\frac{77}{999}$$

❖ Check your answers using the Answer Key at the back of the book.

12 PERCENTAGES

A percentage is an alternative way of expressing a fraction or decimal. We write the percent symbol (%) after a number to indicate a percentage. A percentage is basically a fraction of 100. For example, 29% (read as 29 percent) means $\frac{29}{100}$ and 91% (read as 91%) means $\frac{91}{100}$.

The word "percent" literally means per 100. A percentage is a number that tells you "how many out of 100." For example, if you earn a grade of 97% on an exam, it means that you answered 97 out of every 100 questions correctly.

Have you noticed that we've used the words percent and percentage? There is a difference between them. We use the word percent to represent a specific value, like 5% or 81%. The word percentage refers to a more general amount, like the percentage of students who studied well the night before the test. If you state the specific number, use the word percent instead of percentage, as in "63 percent of the nation." If you don't include a specific number, use the word percentage instead of percent, as in "a percentage of the nation."

Definitions

- **percent**: a specific amount out of 100. For example, 23% means an average of 23 out of every 100. The value 23% is read as "23 percent."
- **percentage**: a general amount out of 100. The word percentage is used instead of percent when a specific value isn't mentioned. For example, compare the general phrase "a percentage of the products" to the specific phrase "18 percent of the products."
- **decimal**: an alternative way to express a fraction, where the denominator is a power of 10. For example, 0.2 means $\frac{2}{10}$ and 0.37 means $\frac{37}{100}$.

To get a better feel for what a percentage means, here are a few common values:

- 100% means all of it. For example, if you score 100% on an exam, it means that you answered every question correctly.
- 50% means one-half. For example, if 50% of the students attended college, it means that half of the students attended college.
- 25% means one-fourth. For example, if 25% of the class brought a calculator, it means that one out of every four students brought a calculator.
- 10% means one-tenth. For example, if you pay a tax of 10% in the store, it means that the tax is one-tenth of the price.
- 200% means double. For example, if stock prices soared to 200% of their original value, it means that the stock prices doubled in value.
- 110% means 10% extra. For example, if you earn a grade of 110% on an assignment, it means that you earned an extra 10% on top of a perfect score.

When reading percentages in sentences, pay close attention to the wording. For example, compare the very similar cases below.

- "Your savings has increased by 50%" means that your savings is 50% greater than it was before. This equates to 50% + 100% = 150%.
- "Your savings has increased to 150%" means the same thing (because 150% means 50% more than 100%) – that your savings is 50% greater than it was before. Note how the little words "by" and "to" make a big difference.
- "Your savings has increased by 150%" is different. This means it is 150% greater than it was before, which is 250% times the original amount.

Converting between decimals and percentages is easy:

- To convert a decimal into a percentage, multiply by 100%.
- To convert a percentage into a decimal, divide by 100%.

Converting between fractions and percentages is similar to Chapters 10 and 11.

- To convert a fraction into a percentage, first convert the fraction into a decimal following the strategy from Chapter 10 or 11 (see page 80 or 96), and then multiply by 100% to convert the decimal into a percentage.
- To convert a percentage into a fraction, first divide by 100% to convert the percentage into a decimal, and then convert the decimal into a fraction following the strategy from Chapter 10 or 11 (see page 78, 88, or 91).

Example: Convert the decimal 0.3 into a percentage.
Simply multiply by 100% (or shift the decimal point 2 places to the right):
$$0.3 \times 100\% = 30\%$$
The decimal 0.3 is equivalent to 30%. We will discuss how to multiply decimals in Chapter 14. For now, if you multiply by a power of 10, shift the decimal point one place to the right for each 0 in the power of 10. For example, 100 has two 0's.

Example: Convert the decimal 1.25 into a percentage.
Simply multiply by 100% (or shift the decimal point 2 places to the right):
$$1.25 \times 100\% = 125\%$$
The decimal 1.25 is equivalent to 125%.

Example: Convert the whole number 3 into a percentage.
Simply multiply by 100% (or shift the decimal point 2 places to the right):
$$3 \times 100\% = 300\%$$
The whole number 3 is equivalent to 300%.

Example: Convert 40% into a decimal.
Simply divide by 100% (or shift the decimal point 2 places to the left):
$$40\% \div 100\% = 0.4$$
The value 40% is equivalent to the decimal 0.4 (note that it's customary to omit the trailing zero from a decimal: that is, 0.40 = 0.4). We will discuss how to divide decimals in Chapter 14. For now, if you divide by a power of 10, shift the decimal point one place to the left for each 0 in the power of 10. For example, 100 has two 0's.

Example: Convert 130% into a decimal.
Simply divide by 100% (or shift the decimal point 2 places to the left):
$$130\% \div 100\% = 1.3$$
The value 130% is equivalent to the decimal 1.3.

Example: Convert 400% into a decimal.
Simply divide by 100% (or shift the decimal point 2 places to the left):
$$400\% \div 100\% = 4$$
The value 400% is equivalent to the whole number 4.

Example: Convert the fraction $\frac{3}{5}$ into a percentage.

- First convert the fraction $\frac{3}{5}$ into a decimal using the method from page 80.
  - The denominator (5) evenly divides into 10 since $5 \times 2 = 10$.
  - Multiply both the numerator and denominator of $\frac{3}{5}$ by 2 in order to make an equivalent fraction with a denominator of 10.
  $$\frac{3}{5} = \frac{3 \times 2}{5 \times 2} = \frac{6}{10}$$
  - Write the new fraction $\frac{6}{10}$ as a decimal with the final digit of the new numerator (6) in the tenths place (corresponding to $\frac{1}{10}$).
  - The fraction $\frac{3}{5}$ is equivalent to the decimal 0.6 (since $0.6 = \frac{6}{10} = \frac{3}{5}$).
- Now multiply by 100% (or shift the decimal point 2 places to the right) to convert the decimal 0.6 into a percentage.
$$0.6 \times 100\% = 60\%$$

The fraction $\frac{3}{5}$ is equivalent to 60% (since $\frac{3}{5} = \frac{6}{10} = 0.6 = 60\%$).

Example: Convert the fraction $\frac{7}{20}$ into a percentage.

- First convert the fraction $\frac{7}{20}$ into a decimal using the method from page 80.
  - The denominator (20) evenly divides into 100 since $20 \times 5 = 100$.
  - Multiply both the numerator and denominator of $\frac{7}{20}$ by 5 in order to make an equivalent fraction with a denominator of 100.
  $$\frac{7}{20} = \frac{7 \times 5}{20 \times 5} = \frac{35}{100}$$
  - Write the new fraction $\frac{35}{100}$ as a decimal with the final digit of the new numerator (35) in the hundredths place (corresponding to $\frac{1}{100}$).
  - The fraction $\frac{7}{20}$ is equivalent to the decimal 0.35 (since $0.35 = \frac{35}{100} = \frac{7}{20}$).
- Now multiply by 100% (or shift the decimal point 2 places to the right) to convert the decimal 0.35 into a percentage.
$$0.35 \times 100\% = 35\%$$

The fraction $\frac{7}{20}$ is equivalent to 35% (since $\frac{7}{20} = \frac{35}{100} = 0.35 = 35\%$).

Example: Convert the fraction $\frac{5}{2}$ into a percentage.

- First convert the fraction $\frac{5}{2}$ into a decimal using the method from page 80.
  - The denominator (2) evenly divides into 10 since $2 \times 5 = 10$.
  - Multiply both the numerator and denominator of $\frac{5}{2}$ by 5 in order to make an equivalent fraction with a denominator of 10.
$$\frac{5}{2} = \frac{5 \times 5}{2 \times 5} = \frac{25}{10}$$
  - Write the new fraction $\frac{25}{10}$ as a decimal with the final digit of the new numerator (25) in the tenths place (corresponding to $\frac{1}{10}$).
  - The fraction $\frac{25}{10}$ is equivalent to the decimal 2.5 (since $2.5 = \frac{25}{10} = \frac{5}{2}$).
- Now multiply by 100% (or shift the decimal point 2 places to the right) to convert the decimal 2.5 into a percentage.
$$2.5 \times 100\% = 250\%$$

The fraction $\frac{5}{2}$ is equivalent to 250% (since $\frac{5}{2} = \frac{25}{10} = 2.5 = 250\%$).

Notes: The final digit of 25, which is a 5, is in the tenths place in 2.5 because the denominator of $\frac{25}{10}$ is 10. The answer (250%) is bigger than 100% and the decimal (2.5) is bigger than 1 because the fraction $\left(\frac{5}{2}\right)$ is bigger than 1.

Example: Convert 175% into a fraction.

- First divide by 100% (or shift the decimal point 2 places to the left) to convert 175% into a decimal.
$$175\% \div 100\% = 1.75$$
- The value 175% is equivalent to the decimal 1.75.
- Next convert the decimal 1.75 into a fraction using the method from page 78.
  - Remove the decimal point from 1.75 to get 175.
  - The final digit (5) of 1.75 is in the hundredths place. Therefore, we divide 175 by one hundred to get $\frac{175}{100}$.
  - The answer can be reduced since 175 and 100 are both divisible by 25. Divide the numerator and denominator both by 25.
$$\frac{175}{100} = \frac{175 \div 25}{100 \div 25} = \frac{7}{4}$$

The value 175% is equivalent to the fraction $\frac{7}{4}$ (since $175\% = 1.75 = \frac{175}{100} = \frac{7}{4}$).

Example: Convert 20% into a fraction.

- First divide by 100% (or shift the decimal point 2 places to the left) to convert 20% into a decimal.

$$20\% \div 100\% = 0.2$$

- The value 20% is equivalent to the decimal 0.2.
- Next convert the decimal 0.2 into a fraction using the method from page 78.
  - Remove the decimal point from 0.2 to get 2. (Ignore the leading zero.)
  - The final digit (2) of 0.2 is in the tenths place. Therefore, we divide 2 by ten to get $\frac{2}{10}$.
  - The answer can be reduced since 2 and 10 are both divisible by 2. Divide the numerator and denominator both by 2.

$$\frac{2}{10} = \frac{2 \div 2}{10 \div 2} = \frac{1}{5}$$

The value 20% is equivalent to the fraction $\frac{1}{5}$ (since $20\% = 0.2 = \frac{2}{10} = \frac{1}{5}$).

Example: Convert 4% into a fraction.

- First divide by 100% (or shift the decimal point 2 places to the left) to convert 4% into a decimal.

$$4\% \div 100\% = 0.04$$

- The value 4% is equivalent to the decimal 0.04. Since 4% means 4 out of 100, it makes sense that the 4 is in the hundredths place in the decimal 0.04.
- Next convert the decimal 0.04 into a fraction using the method from page 78.
  - Remove the decimal point from 0.04 to get 4.
  - The final digit (4) of 0.04 is in the hundredths place. Therefore, we divide 4 by one hundred to get $\frac{4}{100}$.
  - The answer can be reduced since 4 and 100 are both divisible by 4. Divide the numerator and denominator both by 4.

$$\frac{4}{100} = \frac{4 \div 4}{100 \div 4} = \frac{1}{25}$$

The value 4% is equivalent to the fraction $\frac{1}{25}$ (since $4\% = 0.04 = \frac{4}{100} = \frac{1}{25}$).

Alternate method: Since a percentage is out of 100, we can simply divide the given percentage by 100. That is, we can start with $4\% = \frac{4}{100}$ and then reduce this to $\frac{1}{25}$.

Date: _____   Score: ___/5   Name: _____

Chapter 12 Exercises – Part A

Directions: Convert each decimal into a percentage.

❶

0.7

❷

0.45

❸

1.1

❹

0.005

❺

2

❖ Check your answers using the Answer Key at the back of the book.

Date: _____　　　Score: ___/5　　　Name: _____

Chapter 12 Exercises – Part B

Directions: Convert each percentage into a decimal.

❻

　　20%

❼

　　500%

❽

　　2.5%

❾

　　325%

❿

　　0.2%

❖ Check your answers using the Answer Key at the back of the book.

Date: _____   Score: ___/5   Name: _____

Chapter 12 Exercises – Part C

Directions: Convert each fraction into a percentage.

⑪

$$\frac{3}{4}$$

⑫

$$\frac{3}{5}$$

⑬

$$\frac{8}{25}$$

⑭

$$\frac{3}{200}$$

⑮

$$\frac{5}{4}$$

❖ Check your answers using the Answer Key at the back of the book.

Date: _____   Score: ___/5   Name: _____

Chapter 12 Exercises – Part D

Directions: Convert each percentage into a reduced fraction.

⓰

50%

⓱

85%

⓲

150%

⓳

25%

⓴

8%

❖ Check your answers using the Answer Key at the back of the book.

13 ADDING AND SUBTRACTING DECIMALS

One way to add or subtract decimals is to follow these steps:
- Align the decimals in a vertical column. The two decimals must line up at the decimal point. (Don't align the right edge like you would for whole numbers.)
- Does one decimal have more decimal places than the other? If so, you can add trailing zeros to the number with fewer decimal places so that both numbers have the same number of decimal places. This step is optional.
- Once the decimals are aligned at the decimal point, add or subtract the numbers like you would normally add or subtract whole numbers. (If you're not fluent in multi-digit addition and subtraction with whole numbers, it may help to review that before you attempt to add or subtract decimals.)
- Preserve the placement of the decimal point in your final answer. That is, the decimal point of your final answer should line up with the decimal points of the numbers that you are adding or subtracting.

Compare the addition problems shown below.
- The problem on the left shows how we would add 25 and 8 with a vertical stack. These whole numbers are aligned at the right.
- The middle problem shows how to add 4.2 and 0.17. These decimals are aligned at the decimal point.
- We added an optional trailing zero to 4.2 in the problem at the right.

$$
\begin{array}{r} 25 \\ +\ 8 \\ \hline 33 \end{array}
\qquad
\begin{array}{r} 4.2 \\ +\ 0.17 \\ \hline 4.37 \end{array}
\qquad
\begin{array}{r} 4.20 \\ +\ 0.17 \\ \hline 4.37 \end{array}
$$

Definition

- **decimal**: an alternative way to express a fraction, where the denominator is a power of 10. For example, 0.2 means $\frac{2}{10}$ and 0.37 means $\frac{37}{100}$.

Example: Add the decimals 1.65 and 0.2.
- Align the decimals in a vertical column. Line up their decimal points.
- You may add an optional trailing zero at the end of 0.2 (see below).
- Once the decimal points are lined up, add the numbers like usual.
- The decimal point of the final answer lines up with the other decimal points.
- Note that $165 + 20 = 185$. Compare this to $1.65 + 0.20 = 1.85$.

$$\begin{array}{r} 1.65 \\ + \ 0.2 \\ \hline \end{array} \qquad \begin{array}{r} 1.65 \\ + \ 0.20 \\ \hline \end{array} \qquad \begin{array}{r} 1.65 \\ + \ 0.20 \\ \hline 1.85 \end{array}$$

Our final answer is 1.85.

Example: Add the decimals 0.25 and 0.134.
- Align the decimals in a vertical column. Line up their decimal points.
- You may add an optional trailing zero at the end of 0.25 (see below).
- Once the decimal points are lined up, add the numbers like usual.
- The decimal point of the final answer lines up with the other decimal points.
- Note that $250 + 134 = 384$. Compare this to $0.250 + 0.134 = 0.384$.

$$\begin{array}{r} 0.25 \\ + \ 0.134 \\ \hline \end{array} \qquad \begin{array}{r} 0.250 \\ + \ 0.134 \\ \hline \end{array} \qquad \begin{array}{r} 0.250 \\ + \ 0.134 \\ \hline 0.384 \end{array}$$

Our final answer is 0.384.

Example: Add the decimals 2.8 and 2.7.
- Align the decimals in a vertical column. Line up their decimal points.
- No trailing zero is needed since 2.8 and 2.7 both end in the tenths place.
- Once the decimal points are lined up, add the numbers like usual.
- The decimal point of the final answer lines up with the other decimal points.
- Note that $28 + 27 = 55$. Compare this to $2.8 + 2.7 = 5.5$.

$$\begin{array}{r} 2.8 \\ + \ 2.7 \\ \hline \end{array} \qquad \begin{array}{r} 2.8 \\ + \ 2.7 \\ \hline 5.5 \end{array}$$

Our final answer is 5.5. (Note that $0.8 + 0.7 = 1.5$ and that this 1 adds to $2 + 2$.)

Example: Subtract 0.3 from 1.54.
- Align the decimals in a vertical column. Line up their decimal points.
- You may add an optional trailing zero at the end of 0.3 (see below).
- Once the decimal points are lined up, subtract the numbers like usual.
- The decimal point of the final answer lines up with the other decimal points.
- Note that $154 - 30 = 124$. Compare this to $1.54 - 0.30 = 1.24$.

$$\begin{array}{r} 1.54 \\ -\ 0.3 \\ \hline \end{array} \qquad \begin{array}{r} 1.54 \\ -\ 0.30 \\ \hline \end{array} \qquad \begin{array}{r} 1.54 \\ -\ 0.30 \\ \hline 1.24 \end{array}$$

Our final answer is 1.24.

Example: Subtract 0.006 from 0.02.
- Align the decimals in a vertical column. Line up their decimal points.
- You may add an optional trailing zero at the end of 0.02 (see below).
- Once the decimal points are lined up, subtract the numbers like usual.
- The decimal point of the final answer lines up with the other decimal points.
- Note that $20 - 6 = 14$. Compare this to $0.020 - 0.006 = 0.014$.

$$\begin{array}{r} 0.02 \\ -\ 0.006 \\ \hline \end{array} \qquad \begin{array}{r} 0.020 \\ -\ 0.006 \\ \hline \end{array} \qquad \begin{array}{r} 0.020 \\ -\ 0.006 \\ \hline 0.014 \end{array}$$

Our final answer is 0.014. (Note that $10 - 6 = 4$ and this 10 is borrowed from 20.)

Example: Subtract 3.5 from 7.2.
- Align the decimals in a vertical column. Line up their decimal points.
- No trailing zero is needed since 3.5 and 7.2 both end in the tenths place.
- Once the decimal points are lined up, subtract the numbers like usual.
- The decimal point of the final answer lines up with the other decimal points.
- Note that $72 - 35 = 37$. Compare this to $7.2 - 3.5 = 37$.

$$\begin{array}{r} 7.2 \\ -\ 3.5 \\ \hline \end{array} \qquad \begin{array}{r} 7.2 \\ -\ 3.5 \\ \hline 3.7 \end{array}$$

Our final answer is 3.7. (Note that $12 - 5 = 7$ and this 10 is borrowed from 70.)

Date: _____   Score: ___/16   Name: _____

Chapter 13 Exercises

Directions: Solve the following arithmetic problems involving decimals.

❶   5.2
   + 0.6
_____

❷   3.02
   + 0.9
_____

❸   0.06
   + 0.038
_____

❹   0.024
   + 0.018
_____

❺   15.6
   + 8
_____

❻   4.7
   + 5.9
_____

❼   0.8
   + 0.375
_____

❽   0.909
   + 0.001
_____

❾   0.36
   − 0.24
_____

❿   0.854
   − 0.03
_____

⓫   0.089
   − 0.007
_____

⓬   4.2
   − 1.75
_____

⓭   0.5
   − 0.125
_____

⓮   0.072
   − 0.038
_____

⓯   0.762
   − 0.4
_____

⓰   3
   − 0.025
_____

❖ Check your answers using the Answer Key at the back of the book.

14 MULTIPLYING AND DIVIDING DECIMALS

One way to multiply decimals is to follow these steps:
- First ignore the decimal points and multiply the numbers together the same way that you would ordinarily multiply whole numbers. (If you're not fluent in multi-digit multiplication with whole numbers, it may help to review that before you attempt to multiply decimals together.)
- Add up the combined number of decimal places of the two numbers that you are multiplying together. Place a decimal point in the final answer such that the total number of decimal places is the same as the combined number of decimal places of the two numbers that you are multiplying together. If the final answer needs more decimal places than the number of digits it has, you will need to insert one or more zeros after the decimal point. This will be more clear when you read the examples starting on the next page.

One way to divide decimals is to follow these steps:
- Multiply both decimals by the same power of 10 such that both decimals turn into whole numbers. Make sure you multiply each decimal by the same power of 10. For example, to divide $0.02 \div 0.08$, multiply both decimals by 100 to get $2 \div 8$. Note that $0.02 \times 100 = 2$ and $0.08 \times 100 = 8$. (Shift the decimal point one place to the right for each 0 in the power of 10.)
- Next make a fraction using the whole numbers that you found in the previous step. For example, $0.02 \div 0.08$ would become $\frac{2}{8}$.
- Convert the fraction into a decimal using the strategy from Chapter 10.

Definition

- **decimal**: an alternative way to express a fraction, where the denominator is a power of 10. For example, 0.2 means $\frac{2}{10}$ and 0.37 means $\frac{37}{100}$.

Example: Multiply the decimals 0.8 and 0.4.

- First ignore the decimal points and multiply: $8 \times 4 = 32$.
- Now count the decimal places: 0.8 has one decimal place and 0.4 has one decimal place. Combined, there are two decimal places.
- Therefore, our answer must have two decimal places.
- Our final answer is 0.32.

$$\begin{array}{r} 8 \\ \times\,4 \\ \hline 32 \end{array} \qquad \begin{array}{r} 0.8 \\ \times\,0.4 \\ \hline 0.32 \end{array}$$

Example: Multiply the decimals 0.005 and 0.03.

- First ignore the decimal points and multiply: $5 \times 3 = 15$.
- Now count the decimal places: 0.005 has 3 decimal places and 0.03 has 2 decimal places. Combined, there are 5 decimal places (since $3 + 2 = 5$). Be sure to count all digits to the right of the decimal point (not just the zeros).
- Therefore, our answer must have 5 decimal places. We will need to insert 3 zeros before the 15 in order to make a total of 5 decimal places.
- Our final answer is 0.00015.

$$\begin{array}{r} 5 \\ \times\,3 \\ \hline 15 \end{array} \qquad \begin{array}{r} 0.005 \\ \times\,0.03 \\ \hline 0.00015 \end{array}$$

Example: Multiply the decimals 0.25 and 3.

- First ignore the decimal points and multiply: $25 \times 3 = 75$.
- Now count the decimal places: 0.25 has two decimal places and 3 doesn't have any decimal places. Combined, there are two decimal places.
- Therefore, our answer must have two decimal places.
- Our final answer is 0.75.

$$\begin{array}{r} 25 \\ \times\,3 \\ \hline 75 \end{array} \qquad \begin{array}{r} 0.25 \\ \times\,3 \\ \hline 0.75 \end{array}$$

Example: Multiply the decimals 0.06 and 0.9.

- First ignore the decimal points and multiply: $6 \times 9 = 54$.
- Now count the decimal places: 0.06 has two decimal places and 0.9 has one decimal place. Combined, there are three decimal places.
- Therefore, our answer must have three decimal places. We will need to insert a zero before the 54 in order to make a total of three decimal places.
- Our final answer is 0.054.

$$
\begin{array}{r} 6 \\ \times\,9 \\ \hline 54 \end{array}
\qquad
\begin{array}{r} 0.06 \\ \times\,0.9 \\ \hline 0.054 \end{array}
$$

Example: Multiply the decimals 0.12 and 0.11.

- First ignore the decimal points and multiply: $12 \times 11 = 132$.
- Now count the decimal places: 0.12 has 2 decimal places and 0.11 has 2 decimal places. Combined, there are 4 decimal places.
- Therefore, our answer must have 4 decimal places. We will need to insert a zero before the 132 in order to make a total of 4 decimal places.
- Our final answer is 0.0132.

$$
\begin{array}{r} 12 \\ \times\,11 \\ \hline 132 \end{array}
\qquad
\begin{array}{r} 0.12 \\ \times\,0.11 \\ \hline 0.0132 \end{array}
$$

Example: Multiply the decimals 0.00004 and 0.02.

- First ignore the decimal points and multiply: $4 \times 2 = 8$.
- Now count the decimal places: 0.00004 has 5 decimal places and 0.02 has 2 decimal places. Combined, there are 7 decimal places (since $5 + 2 = 7$). Be sure to count all digits to the right of the decimal point (not just the zeros).
- Therefore, our answer must have 7 decimal places. We will need to insert 6 zeros before the 8 in order to make a total of 7 decimal places.
- Our final answer is 0.0000008.

$$
\begin{array}{r} 4 \\ \times\,2 \\ \hline 8 \end{array}
\qquad
\begin{array}{r} 0.00004 \\ \times\,0.02 \\ \hline 0.0000008 \end{array}
$$

Example: Perform the division $0.18 \div 0.12$.

- Note that 0.18 has 2 decimal places and 0.12 also has 2 decimal places.
- Multiply 0.18 and 0.12 both by 100. Shift each decimal point two places to the right: $0.18 \times 100 = 18$ and $0.12 \times 100 = 12$.
- Use these whole numbers to make a fraction:

$$0.18 \div 0.12 = \frac{18}{12}$$

- Note that $\frac{18}{12}$ can be reduced since 18 and 12 are both divisible by 6.

$$\frac{18}{12} = \frac{18 \div 6}{12 \div 6} = \frac{3}{2}$$

- Convert the fraction $\frac{3}{2}$ into a decimal using the strategy from Chapter 10.
  - The denominator (2) evenly divides into 10 since $2 \times 5 = 10$.
  - Multiply both the numerator and denominator of $\frac{3}{2}$ by 5 in order to make an equivalent fraction with a denominator of 10.

$$\frac{3}{2} = \frac{3 \times 5}{2 \times 5} = \frac{15}{10}$$

  - Write the new fraction $\frac{15}{10}$ as a decimal with the final digit of the new numerator (15) in the tenths place (corresponding to $\frac{1}{10}$): $\frac{15}{10} = 1.5$.

Our final answer is 1.5.

Example: Perform the division $1.234 \div 0.1$.

- Note that 1.234 has the most decimal places: 1.234 has 3 decimal places.
- Multiply 1.234 and 0.1 both by 1000. Shift each decimal point 3 places to the right: $1.234 \times 1000 = 1234$ and $0.1 \times 1000 = 100$. (Shift 0.1 one place to the right to make 1 and shift 1 two more places to the right to make 100.)
- Use these whole numbers to make a fraction:

$$1.234 \div 0.1 = \frac{1234}{100}$$

- This example has a shorter solution because $\frac{1234}{100}$ is already over a power of 10. Write the new fraction $\frac{1234}{100}$ as a decimal with the final digit of the new numerator (1234) in the hundredths place (corresponding to $\frac{1}{100}$): $\frac{1234}{100} = 12.34$ (shift the decimal point of 1234 two places to the left to get 12.34).

Our final answer is 12.34.

Example: Perform the division $3 \div 0.4$.

- Note that 0.4 has the most decimal places: 0.4 has one decimal place.
- Multiply 3 and 0.4 both by 10. Shift each decimal point one place to the right: $3 \times 10 = 30$ and $0.4 \times 10 = 4$.
- Use these whole numbers to make a fraction:
$$3 \div 0.4 = \frac{30}{4}$$
- Convert the fraction $\frac{30}{4}$ into a decimal using the strategy from Chapter 10.
  - The denominator (4) evenly divides into 100 since $4 \times 25 = 100$.
  - Multiply both the numerator and denominator of $\frac{30}{4}$ by 25 in order to make an equivalent fraction with a denominator of 100.
$$\frac{30}{4} = \frac{30 \times 25}{4 \times 25} = \frac{750}{100}$$
  - Write the new fraction $\frac{750}{100}$ as a decimal with the final digit of the new numerator (750) in the hundredths place (corresponding to $\frac{1}{100}$):
$\frac{750}{100} = 7.50$. It's customary to ignore the trailing zero: $7.50 = 7.5$.

Our final answer is 7.5.

Example: Perform the division $0.12 \div 0.03$.

- Note that 0.12 has 2 decimal places and 0.03 also has 2 decimal places.
- Multiply 0.12 and 0.03 both by 100. Shift each decimal point 2 places to the right: $0.12 \times 100 = 12$ and $0.03 \times 100 = 3$.
- Use these whole numbers to make a fraction:
$$0.12 \div 0.03 = \frac{12}{3}$$
- Note that $\frac{12}{3}$ can be reduced since 12 and 3 are both divisible by 3.
$$\frac{12}{3} = \frac{12 \div 3}{3 \div 3} = \frac{4}{1} = 4$$
- Note that $\frac{4}{1} = 4 \div 1 = 4$ (or go back a step and note that $\frac{12}{3} = 12 \div 3 = 4$).
- This example has a shorter solution because the answer is a whole number.

Our final answer is 4.

Example: Perform the division $0.0032 \div 0.005$.
- Note that 0.0032 has the most decimal places: 0.0032 has 4 decimal places.
- Multiply 0.0032 and 0.005 both by 10,000. Shift each decimal point 4 places to the right: $0.0032 \times 10{,}000 = 32$ and $0.005 \times 10{,}000 = 50$. (Shift 0.005 three places to the right to make 5 and shift 5 one more place to the right to make 50.)
- Use these whole numbers to make a fraction:
$$0.0032 \div 0.005 = \frac{32}{50}$$
- Convert the fraction $\frac{32}{50}$ into a decimal using the strategy from Chapter 10.
    - The denominator (50) evenly divides into 100 since $50 \times 2 = 100$.
    - Multiply both the numerator and denominator of $\frac{32}{50}$ by 2 in order to make an equivalent fraction with a denominator of 100.
$$\frac{32}{50} = \frac{32 \times 2}{50 \times 2} = \frac{64}{100}$$
    - Write the new fraction $\frac{64}{100}$ as a decimal with the final digit of the new numerator (64) in the hundredths place (corresponding to $\frac{1}{100}$):
$\frac{64}{100} = 0.64$.

Our final answer is 0.64.

Example: Perform the division $2 \div 0.5$.
- Note that 0.5 has the most decimal places: 0.5 has one decimal place.
- Multiply 2 and 0.5 both by 10. Shift each decimal point one place to the right: $2 \times 10 = 20$ and $0.5 \times 10 = 5$.
- Use these whole numbers to make a fraction:
$$2 \div 0.5 = \frac{20}{5}$$
- Note that $\frac{20}{5}$ can be reduced since 20 and 5 are both divisible by 5.
$$\frac{20}{5} = \frac{20 \div 5}{5 \div 5} = \frac{4}{1} = 4$$
- Note that $\frac{4}{1} = 4 \div 1 = 4$ (or go back a step and note that $\frac{20}{5} = 20 \div 5 = 4$).
- This example has a shorter solution because the answer is a whole number.

Our final answer is 4.

Example: Perform the division $0.008 \div 0.02$.

- Note that 0.008 has the most decimal places: 0.008 has 3 decimal places.
- Multiply 0.008 and 0.02 both by 1000. Shift each decimal point 3 places to the right: $0.008 \times 1000 = 8$ and $0.02 \times 1000 = 20$. (Shift 0.02 two places to the right to make 2 and shift 2 one more place to the right to make 20.)
- Use these whole numbers to make a fraction:

$$0.008 \div 0.02 = \frac{8}{20}$$

- Convert the fraction $\frac{8}{20}$ into a decimal using the strategy from Chapter 10.
  - The denominator (20) evenly divides into 100 since $20 \times 5 = 100$.
  - Multiply both the numerator and denominator of $\frac{8}{20}$ by 5 in order to make an equivalent fraction with a denominator of 100.

$$\frac{8}{20} = \frac{8 \times 5}{20 \times 5} = \frac{40}{100}$$

  - Write the new fraction $\frac{40}{100}$ as a decimal with the final digit of the new numerator (40) in the hundredths place (corresponding to $\frac{1}{100}$): $\frac{40}{100} = 0.40$. It's customary to ignore the trailing zero: $0.40 = 0.4$.

Our final answer is 0.4.

Date: _____   Score: ___/12   Name: _____

Chapter 14 Exercises – Part A

Directions: Multiply the given decimals.

❶   0.9
    × 0.3

❷   1.2
    × 0.4

❸   0.05
    × 0.6

❹   0.003
    × 0.02

❺   1.2
    × 1.2

❻   1.75
    × 4

❼   2.25
    × 0.5

❽   0.09
    × 0.04

❾   100
    × 0.7

❿   0.025
    × 0.08

⓫   3000
    × 0.04

⓬   0.0001
    × 0.001

❖ Check your answers using the Answer Key at the back of the book.

Date: _____   Score: ___/4   Name: _____

Chapter 14 Exercises – Part B

Directions: Divide the given decimals.

⑬

$0.48 \div 0.12 =$

⑭

$0.6 \div 0.8 =$

⑮

$4.5 \div 0.01 =$

⑯

$0.75 \div 0.5 =$

❖ Check your answers using the Answer Key at the back of the book.

Date: _____   Score: ___/4   Name: _____

Chapter 14 Exercises – Part C

Directions: Divide the given decimals.

⑰

$0.08 \div 0.4 =$

⑱

$1 \div 0.25 =$

⑲

$0.048 \div 0.32 =$

⑳

$0.00032 \div 0.0032 =$

❖ Check your answers using the Answer Key at the back of the book.

15 WORD PROBLEMS

When you read a word problem, try to identify the following information:
- What do the given numbers represent?
- What are you solving for?
- How are the given numbers related to what you're solving for?

The word problems in this chapter involve fractions, decimals, or percentages. Some problems give you a fraction, decimal, or percentage directly, as in these examples:
- 2/3 of the apples are green.
- The price is $1.75.
- 25% of the products.

The information about a fraction may be given indirectly, as in these examples.
- 3 out of 4 students.
- A ratio of 5 to 2.
- 4 days per week.

Sometimes you're solving for a fraction, decimal, or percentage, as in these examples.
- What fraction of the socks are red?
- How much does each item cost?
- Express your answer as a percentage.
- Determine the ratio of the boys to the girls.

The word problems in this chapter involve techniques that are taught in this book.
- What is a fraction (Chapter 1)?
- Reduce fractions (Chapter 2).
- Compare fractions (Chapter 4).
- Add, subtract, multiply, or divide fractions (Chapters 5–9).
- Add, subtract, multiply, or divide decimals (Chapters 13–14).
- Convert a fraction into a decimal or percentage (Chapters 10–12).
- Convert a decimal into a fraction or percentage (Chapters 10–12).
- Convert a percentage into a fraction or decimal (Chapters 10–12).

The simplest problems in this chapter involve relating a fraction to its numerator and denominator. For example, a problem may tell you that 7 out of 12 cars are red, and ask you to find the fraction of the cars that are red. In this example, 7 is the numerator, 12 is the denominator, and the fraction is $\frac{7}{12}$.

Beware that the numerator isn't always stated before the denominator. In both of the examples that follow, the denominator is the total number of students.
- 71 out of 100 students attended the seminar. In this case, the numerator (71) is stated before the denominator (100). The fraction of students that attended the seminar is $\frac{71}{100}$.
- There are 100 students in the school. 71 of the students attended the seminar. In this case, the denominator (100) is stated before the numerator (71). The fraction of students that attended the seminar is $\frac{71}{100}$.

Beware that some problems don't give you the denominator directly. Compare:
- There are 11 girls in a class of 20 students. The denominator is 20. The fraction of students that are girls is $\frac{11}{20}$.
- There are 9 boys and 11 girls in a class. To find the fraction of students that are girls, you must add $9 + 11 = 20$ to get the denominator. The fraction of students that are girls is $\frac{11}{20}$.

Beware that a "ratio" often doesn't state the denominator directly.
- The ratio of pens to pencils is 7 to 3. To find the fraction of writing utensils that are pens, you must add $7 + 3 = 10$ to get the denominator. The fraction of writing utensils that are pens is $\frac{7}{10}$.
- There are 7 pens for every 3 pencils. This doesn't mention the word "ratio," but it is the same ratio as stated in the previous example.

Beware that numbers may be stated as words.
- half $= \frac{1}{2}$, third $= \frac{1}{3}$, quarter $= \frac{1}{4}$, two-thirds $= \frac{2}{3}$, three quarters $= \frac{3}{4}$, etc.
- words ending in "th": fourth, fifth, sixth, seventh, eighth, ninth, tenth, etc.
- multiples: double, twice, triple, quadruple, five times, six times, etc.

Problems with money may involve decimals.
- The hamburger costs $3.75 (in United States dollars).
- Other countries often use a different symbol (such as £) for currency.

When reading percentages in sentences, pay close attention to the wording. For example, compare the very similar cases below.

- "Your savings has increased by 50%" means that your savings is 50% greater than it was before. This equates to $50\% + 100\% = 150\%$.
- "Your savings has increased to 150%" means the same thing (because 150% means 50% more than 100%) – that your savings is 50% greater than it was before. Note how the little words "by" and "to" make a big difference.
- "Your savings has increased by 150%" is different. This means it is 150% greater than it was before, which is 250% times the original amount.

If a problem involves two fractions or two decimals, you may need to add, subtract, multiply, or divide them. The following keywords can help you determine this.

- Addition: total of, sum, increased by, gained, raised to, more than, greater than, combined, together.
- Subtraction: difference, between, left over, fewer, minus, decreased by, lost, less than, smaller than, taken away, after.
- Multiplication: times, product, increased by a factor of, decreased by a factor of, twice, double, triple, each, of.
- Division: out of, per, ratio, split, equal pieces, average.

Beware that there are many different ways to use language to convey the same idea, and that language can sometimes use a word in an unusual way. Therefore, the above keywords are not foolproof. They can be helpful, but more important than the keywords is the ability to reason out what the problem is saying. Practice helps, but you must actively think your way through the examples and problems.

Beware that a rare problem may include irrelevant information. In most problems, you must use all of the given information. The word problems in this chapter use all the known quantities. However, beware that there are a few tricky word problems in some books that include information that is irrelevant to the solution.

Definitions

- **numerator**: the number that appears in the top of a fraction.
- **denominator**: the number that appears in the bottom of a fraction.
- **percent**: a specific amount out of 100, such as 23% of the students.
- **percentage**: a general amount out of 100, such as a percentage of the students.

Example: 8 out of 12 monkeys have bananas. What fraction of the monkeys have bananas?

- 8 monkeys have bananas. The numerator is 8.
- There are 12 monkeys all together. The denominator is 12.
- Divide the numerator by the denominator to find the fraction of monkeys that have bananas. The fraction is $\frac{8}{12}$.
- The fraction $\frac{8}{12}$ can be reduced. Divide the numerator and denominator both by 4 in order to reduce the fraction.

$$\frac{8}{12} = \frac{8 \div 4}{12 \div 4} = \frac{2}{3}$$

The fraction of the monkeys that have bananas is $\frac{2}{3}$.

Example: 6 monkeys have bananas, while 4 other monkeys have coconuts. What fraction of the monkeys have bananas?

- 6 monkeys have bananas. The numerator is 6.
- Add $6 + 4 = 10$ to find the total number of monkeys. The denominator is 10.
- Divide the numerator by the denominator to find the fraction of monkeys that have bananas. The fraction is $\frac{6}{10}$.
- The fraction $\frac{6}{10}$ can be reduced. Divide the numerator and denominator both by 2 in order to reduce the fraction.

$$\frac{6}{10} = \frac{6 \div 2}{10 \div 2} = \frac{3}{5}$$

The fraction of the monkeys that have bananas is $\frac{3}{5}$.

Example: 5 monkeys have apples, while 3 other monkeys have oranges. What fraction of the monkeys have oranges?

- 3 monkeys have oranges. The numerator is 3. The question asks about oranges, so the numerator is 3 (not 5).
- Add $5 + 3 = 8$ to find the total number of monkeys. The denominator is 8.
- Divide the numerator by the denominator to find the fraction of monkeys that have oranges. The fraction is $\frac{3}{8}$.

The fraction of the monkeys that have oranges is $\frac{3}{8}$.

Example: The ratio of monkeys to gorillas is 15 to 6. What fraction of these animals are monkeys?

- 15 of the animals are monkeys. The numerator is 15.
- Add $15 + 6 = 21$ to find the total number of animals. The denominator is 21.
- Divide the numerator by the denominator to find the fraction of the animals that are monkeys. The fraction is $\frac{15}{21}$.
- The fraction $\frac{15}{21}$ can be reduced. Divide the numerator and denominator both by 3 in order to reduce the fraction.

$$\frac{15}{21} = \frac{15 \div 3}{21 \div 3} = \frac{5}{7}$$

The fraction of the animals that are monkeys is $\frac{5}{7}$.

Example: 10 out of 20 monkeys have bananas. What percentage of the monkeys have bananas?

- 10 monkeys have bananas. The numerator is 10.
- There are 20 monkeys all together. The denominator is 20.
- Divide the numerator by the denominator to find the fraction of monkeys that have bananas. The fraction of the monkeys that have bananas is $\frac{10}{20}$.
- The question asked for a percentage, not a fraction. Convert the fraction to a percentage using the strategy from Chapter 12.
  - The denominator (20) evenly divides into 100 since $20 \times 5 = 100$.
  - Multiply both the numerator and denominator of $\frac{10}{20}$ by 5 in order to make an equivalent fraction with a denominator of 100.

$$\frac{10}{20} = \frac{10 \times 5}{20 \times 5} = \frac{50}{100}$$

  - Write the new fraction $\frac{50}{100}$ as a decimal with the final digit of the new numerator (50) in the hundredths place (corresponding to $\frac{1}{100}$).
  - The fraction $\frac{10}{20}$ is equivalent to the decimal 0.5 (since $0.5 = 0.50 = \frac{50}{100} = \frac{10}{20}$).
- Now multiply by 100% (or shift the decimal point 2 places to the right) to convert the decimal 0.5 into a percentage.

$$0.5 \times 100\% = 50\%$$

The percentage of the monkeys that have bananas is 50%.

Example: A monkey ate 2/5 of his bananas for breakfast and ate 1/4 of his bananas for lunch. What fraction of his bananas did the monkey eat all together?

- The words "all together" can help you realize that you need to add these fractions. Apply the strategy from Chapter 5 to add the fractions.

- Multiply the numerator and denominator of $\frac{2}{5}$ by 4 (because $5 \times 4 = 20$) in order to make an equivalent fraction with a denominator of 20.

$$\frac{2}{5} = \frac{2 \times 4}{5 \times 4} = \frac{8}{20}$$

- Multiply the numerator and denominator of $\frac{1}{4}$ by 5 (because $4 \times 5 = 20$) in order to make an equivalent fraction with a denominator of 20.

$$\frac{1}{4} = \frac{1 \times 5}{4 \times 5} = \frac{5}{20}$$

- Now we can express the problem with common denominators $\left(\frac{8}{20} \text{ and } \frac{5}{20}\right)$, since $\frac{2}{5} = \frac{8}{20}$ and $\frac{1}{4} = \frac{5}{20}$.

$$\frac{2}{5} + \frac{1}{4} = \frac{8}{20} + \frac{5}{20} = \frac{8+5}{20} = \frac{13}{20}$$

The monkey ate $\frac{13}{20}$ of his bananas all together.

Example: A monkey ate 1/4 of his bananas for breakfast. What fraction of his bananas does the monkey have left over?

- The words "left over" can help you realize that you need to subtract these fractions. Apply the strategy from Chapter 6 to subtract the fractions.

- Subtract $\frac{1}{4}$ from 1. Why 1? It's because 1 represents the whole amount of bananas that the monkey had originally.

- Write 1 as $1 = \frac{1}{1} = \frac{4}{4}$ to make a common denominator of 4.

$$1 - \frac{1}{4} = \frac{1}{1} - \frac{1}{4} = \frac{1 \times 4}{1 \times 4} - \frac{1}{4} = \frac{4}{4} - \frac{1}{4} = \frac{4-1}{4} = \frac{3}{4}$$

The monkey has $\frac{3}{4}$ of his bananas left over.

Note: The above example could have asked, "What fraction of the monkey's bananas remain?" Don't count on the problem saying "left over." There are many different ways to use language to communicate the same idea.

Example: A monkey had 8 bananas. The monkey ate 3/4 of his bananas. How many bananas did the monkey eat?

- Note: This problem gave us the fraction. We're not solving for a fraction.
- The word "of" sometimes refers to multiplication. This is one of the times: We can determine "$\frac{3}{4}$ of his bananas" by multiplying $\frac{3}{4} \times 8$.
- Apply the strategy from Chapter 7 to multiply fractions. Note that $8 = \frac{8}{1}$.

$$\frac{3}{4} \times 8 = \frac{3}{4} \times \frac{8}{1} = \frac{3 \times 8}{4 \times 1} = \frac{24}{4}$$

- The fraction $\frac{24}{4}$ can be reduced. Divide the numerator and denominator both by 4 in order to reduce the fraction.

$$\frac{24}{4} = \frac{24 \div 4}{4 \div 4} = \frac{6}{1}$$

- Note that $\frac{6}{1} = 6 \div 1 = 6$ or go back a step and note that $\frac{24}{4} = 24 \div 4 = 6$.

The monkey ate 6 of his bananas.

Example: A monkey had 5 bananas. The monkey ate 40% of his bananas. How many bananas did the monkey eat?

- This problem gave us the percentage. We're not solving for a fraction in this example. We're solving for the number of bananas that the monkey ate.
- The word "of" sometimes refers to multiplication. This is one of the times: We can determine "40% of his bananas" by multiplying 40% × 5.
- First convert the percentage to a decimal. Divide 40% by 100% (or shift the decimal two places to the left): 40% = 0.40, which is the same as 0.4.
- Apply the strategy from Chapter 14 to multiply decimals: 0.4 × 5 = 2.

The monkey ate 2 of his bananas.

Example: A banana costs $1.25 and a watermelon costs $3.50. What would the total cost be for both items?

- The words "total cost" can help you realize that you need to add these decimals. Apply the strategy from Chapter 13 to add the decimals.

$$\begin{array}{r} \$3.50 \\ + \ \$1.25 \\ \hline \$4.75 \end{array}$$

The total cost for both items is $4.75.

Example: A coconut costs $0.75. A monkey has a $1 bill. How much change will the monkey receive?

- This problem doesn't have the usual keywords to indicate which type of problem it is. However, note that the change is the money that is "left over."
- Apply the strategy from Chapter 13 to subtract the decimals.

$$\begin{array}{r} \$1.00 \\ -\ \$0.75 \\ \hline \$0.25 \end{array}$$

The monkey will receive $0.25 in change.

Example: Apples cost $0.42 each. How much would it cost to buy three apples?

- The word "each" sometimes indicates multiplication. In this example, it will cost "three times" as much to buy three apples, which requires multiplying $0.42 × 3. Apply the strategy from Chapter 14 to multiply decimals.

$$\begin{array}{r} \$0.42 \\ \times\ 3 \\ \hline \$1.26 \end{array}$$

It would cost $1.26 to buy three apples.

Example: A monkey and a chimpanzee discovered a pile of bananas. The monkey ate 2/5 of the bananas and the chimpanzee at 3/8 of the bananas. Who ate more of the bananas?

- This is a comparison problem. We need to determine whether 2/5 is larger, smaller than, or equal to 3/8.
- Apply the strategy from Chapter 4 to compare two fractions. We will apply the alternate method (cross multiplying) from Chapter 4.
- Cross multiply the fractions $\frac{2}{5}$ and $\frac{3}{8}$ as follows:
  - Multiply the left numerator (2) by the right denominator (8). Write this on the left.
  - Multiply the right numerator (3) by the left denominator (5). Write this on the right.
  - We get 2 × 8 and 3 × 5. This simplifies to 16 and 15.
  - Since 16 > 15, it follows that $\frac{2}{5} > \frac{3}{8}$ (meaning that $\frac{2}{5}$ is greater than $\frac{3}{8}$).

The monkey ate more bananas than the chimpanzee ate.

Example: A monkey pays \$2.50 to buy 5 mangos. How much does each mango cost?

- This example uses the word "each," yet this example involves division rather than multiplication. We must divide the total cost by 5 in order to find the cost of one mango. Apply the strategy from Chapter 14 to find \$2.50 ÷ 5.
- Note that 2.50 has the most decimal places: 2.50 has 2 decimal places.
- Multiply 2.50 and 5 both by 100. Shift each decimal point two places to the right: $2.50 \times 100 = 250$ and $5 \times 100 = 500$.
- Use these whole numbers to make a fraction:

$$2.50 \div 5 = \frac{250}{500}$$

- Convert the fraction $\frac{250}{500}$ into a decimal using the strategy from Chapter 10.
  - The denominator (500) evenly divides into 1000 since $500 \times 2 = 1000$.
  - Multiply both the numerator and denominator of $\frac{250}{500}$ by 2 in order to make an equivalent fraction with a denominator of 1000.

$$\frac{250}{500} = \frac{250 \times 2}{500 \times 2} = \frac{500}{1000}$$

  - Write the new fraction $\frac{500}{1000}$ as a decimal with the final digit of the new numerator (the last 0) in the thousandths place (corresponding to $\frac{1}{1000}$): $\frac{500}{1000} = 0.500$. This is the same as 0.50 or 0.5 (or .5).

The mangos cost \$0.50 each (assuming that the monkey didn't receive any discount for buying multiple mangos).

Example: A shirt costs \$30. How much would you save with a 20% off coupon?

- The savings will equal 20% of \$30. We need to multiply $20\% \times \$30$.
- First convert the percentage to a decimal. Divide 20% by 100% (or shift the decimal two places to the left): $20\% = 0.20$, which is the same as 0.2.
- Apply the strategy from Chapter 14 to multiply decimals: $0.2 \times \$30 = \$6$.

You would save \$6.

Note: The previous example asked about the savings (or discount). If instead the question asked how much it would cost to buy the shirt, the answer would be different.

Example: A melon costs $1.50. What would the total be if there is a 10% sales tax?
- First determine the sales tax. The sales tax equals 10% of $1.50. We need to multiply 10% × $1.50.
- Next convert the percentage to a decimal. Divide 10% by 100% (or shift the decimal two places to the left): 10% = 0.10, which is the same as 0.1.
- Apply the strategy from Chapter 14 to multiply decimals:

$$\begin{array}{r} \$1.50 \\ \times \ \ 0.1 \\ \hline \$0.15 \end{array}$$

- Add the cost of the melon ($1.50) to the sales tax ($0.15) to determine the total cost:

$$\begin{array}{r} \$1.50 \\ + \ \ \$0.15 \\ \hline \$1.65 \end{array}$$

The total would be $1.65.

Example: A monkey has $24 in her savings account. If her savings increases by 50%, how much money will she have in her savings account then?
- Read this question carefully: It doesn't ask you to find 50% of $24. Rather, it says that $24 increases by 50%. This means to first find 50% of $24 and then add that amount to $24. (Alternatively, you could find 150% of $24.)
- First multiply 50% × $24. (This won't be the final answer, but we will use this as an intermediate answer.)
- Convert the percentage to a decimal. Divide 50% by 100% (or shift the decimal two places to the left): 50% = 0.50, which is the same as 0.5.
- Apply the strategy from Chapter 14 to multiply decimals: 0.5 × $24 = $12.
- Add the increased amount ($12) that we just found to the original savings ($24) to determine how much she will have in her savings after the increase: $24 + $12 = $36.

She would have $36 in her savings account after it increases by 50%.

Date: _____   Score: ___/20   Name: _____

Chapter 15 Exercises

Directions: Solve the following word problems.

❶ If 16 out of 20 monkeys prefer to peel their own bananas, what fraction of the monkeys prefer to peel their own bananas?

❷ If 6 monkeys are wearing red shirts and 3 monkeys are wearing blue shirts, what fraction of the monkeys are wearing red shirts?

❖ Check your answers using the Answer Key at the back of the book.

❸ If 24 monkeys voted yes and 18 monkeys voted no, what fraction of the monkeys voted no?

❹ The ratio of apples to oranges is 16 to 4. What fraction of these fruits are apples?

❖ Check your answers using the Answer Key at the back of the book.

❺ There are 12 female monkeys and 6 male monkeys in a room. What is the ratio of females to males?

❻ A monkey jogged for 3/8 of an hour in the morning and for 5/6 of an hour in the afternoon. For how much time did the monkey jog all together?

❖ Check your answers using the Answer Key at the back of the book.

❼ A monkey wishes to jog for 3/2 hours today. The monkey already jogged for 5/8 of an hour. For how much more time does the monkey still need to jog?

❽ A pie needs to bake for 7/12 of an hour. The pie has already baked for 2/5 of an hour. For how much more time must the pie bake?

❖ Check your answers using the Answer Key at the back of the book.

❾ A monkey had 30 bananas. The monkey ate 1/6 of his bananas. How many bananas did the monkey eat?

❿ A monkey had 28 bananas. The monkey ate 25% of his bananas. How many bananas did the monkey eat?

❖ Check your answers using the Answer Key at the back of the book.

❶❶ If 3 out of 20 monkeys are taking a nap, what percentage of the monkeys are taking a nap?

❶❷ A monkey traveled a distance of 7/4 kilometers. A gorilla traveled a distance of 8/5 kilometers. Who traveled the greatest distance?

❖ Check your answers using the Answer Key at the back of the book.

⓭ An apple costs $0.38 and an orange costs $0.84. What would the total cost be for both items?

⓮ A watermelon costs $3.25. A monkey pays with a $5 bill. How much money will the monkey receive back?

❖ Check your answers using the Answer Key at the back of the book.

⓯ A store sells bananas for $0.25 each. How much would it cost to buy 9 bananas?

⓰ A monkey buys 8 pears. The total price is $2.40. How much would it cost to buy just one pear?

❖ Check your answers using the Answer Key at the back of the book.

❶⑦ A store sells a jacket for $30. If you have a coupon for a 25% discount, how much money will you save?

❶⑧ A dress costs $45. What would the total be if there is an 8% sales tax?

❖ Check your answers using the Answer Key at the back of the book.

❿ A store sells a suit for $150. You have a coupon for a 30% discount. There is a sales tax of 10%. The discount is applied before the sales tax is applied. What will be the total purchase price?

⓴ A monkey has $360 in her savings account. If her savings increases by 20%, how much money will she have in her savings account then?

❖ Check your answers using the Answer Key at the back of the book.

16 MIXED NUMBERS

A mixed number combines a whole number together with a fraction. For example, the mixed number $2\frac{1}{4}$ represents two and one-fourth. The 2 and the $\frac{1}{4}$ are added together to make $2\frac{1}{4}$.

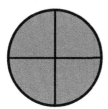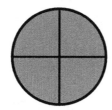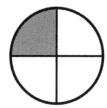

An alternative way to express a mixed number is with an improper fraction. The term improper fraction refers to a fraction where the numerator is larger than the denominator. For example, $\frac{4}{3}$ and $\frac{9}{2}$ are improper fractions.

The name "improper" can be misleading. In many math and science classes, it is actually more common to work with improper fractions than mixed numbers. The word "improper" doesn't mean you should avoid it at all costs.

Sometimes you have a mixed number and it would be convenient to express it as an improper fraction. Other times you have an improper fraction and it would be convenient to express it as a mixed number. In this chapter, we will learn how to convert mixed numbers into improper fractions, and vice-versa.

Definitions

- **mixed number**: a whole number and fraction combined together, such as $5\frac{3}{8}$.
- **improper fraction**: a fraction for which the numerator is larger than the denominator, such as $\frac{43}{8}$.

To convert a mixed number, like $3\frac{1}{4}$, into an improper fraction, follow these steps:
- Multiply the whole number with the denominator.
- Add this to the original numerator in order to make the new numerator.
- Put the new numerator over the original denominator.

Here is a quick example:

$$3\frac{1}{4} = \frac{3 \times 4 + 1}{4} = \frac{12 + 1}{4} = \frac{13}{4}$$

To convert an improper fraction, like $\frac{13}{4}$, into a mixed number, follow these steps:
- Divide the numerator by the denominator.
- Express the answer as an integer plus a remainder. (See below for a brief review of remainders.)
- The mixed number equals the integer plus the remainder over the original denominator.

In the example below, $13 \div 4 = 3$ R1, where 3 R1 means 3 with a remainder of 1.

$$\frac{13}{4} = 13 \div 4 = 3 \text{ R1} = 3 + \frac{1}{4} = 3\frac{1}{4}$$

Remainders

When a whole number doesn't divide evenly into another whole number, we can express the answer using a remainder. Following are some examples.
- $14 \div 3$: What's the largest number that 3 divides into evenly, which isn't bigger than 14? The answer is 4 because $3 \times 4 = 12$. Since $3 \times 4 = 12$, there is a remainder of 2. Therefore, $14 \div 3 = 4$ R2. This means that 14 divided by 3 equals 4 with a remainder of 2. Another way to say this is that if you multiply 3×4, the answer will be 2 less than 14.
- $33 \div 7$: Think of the multiples of 7, which are 7, 14, 21, 28, 35, 42, 49, etc. What is the largest multiple of 7 that isn't bigger than 33? The answer is 28. What must you multiply 7 by to make 28? It's 4 (since $7 \times 4 = 28$). Subtract 28 from 33 to find the remainder: $33 - 28 = 5$. Therefore, $33 \div 7 = 4$ R5.
- $19 \div 8 = 2$ R3 because $8 \times 2 = 16$ and $19 - 16 = 3$.
- $21 \div 5 = 4$ R1 because $5 \times 4 = 20$ and $21 - 20 = 1$.
- $6 \div 4 = 1$ R2 because $4 \times 1 = 4$ and $6 - 4 = 2$.

Example: Convert the mixed number $3\frac{2}{5}$ into an improper fraction.

- Multiply the whole number with the denominator: $3 \times 5 = 15$.
- Add the answer to the previous step to the numerator: $15 + 2 = 17$.
- Put the new numerator over the original denominator: $\frac{17}{5}$.

$$3\frac{2}{5} = \frac{3 \times 5 + 2}{5} = \frac{15 + 2}{5} = \frac{17}{5}$$

The improper fraction $\frac{17}{5}$ is equivalent to the mixed number $3\frac{2}{5}$.

Example: Convert the mixed number $2\frac{3}{4}$ into an improper fraction.

- Multiply the whole number with the denominator: $2 \times 4 = 8$.
- Add the answer to the previous step to the numerator: $8 + 3 = 11$.
- Put the new numerator over the original denominator: $\frac{11}{4}$.

$$2\frac{3}{4} = \frac{2 \times 4 + 3}{4} = \frac{8 + 3}{4} = \frac{11}{4}$$

The improper fraction $\frac{11}{4}$ is equivalent to the mixed number $2\frac{3}{4}$.

Example: Convert the mixed number $9\frac{5}{8}$ into an improper fraction.

- Multiply the whole number with the denominator: $9 \times 8 = 72$.
- Add the answer to the previous step to the numerator: $72 + 5 = 77$.
- Put the new numerator over the original denominator: $\frac{77}{8}$.

$$9\frac{5}{8} = \frac{9 \times 8 + 5}{8} = \frac{72 + 5}{8} = \frac{77}{8}$$

The improper fraction $\frac{77}{8}$ is equivalent to the mixed number $9\frac{5}{8}$.

Example: Convert the mixed number $1\frac{2}{3}$ into an improper fraction.

- Multiply the whole number with the denominator: $1 \times 3 = 3$.
- Add the answer to the previous step to the numerator: $3 + 2 = 5$.
- Put the new numerator over the original denominator: $\frac{5}{3}$.

$$1\frac{2}{3} = \frac{1 \times 3 + 2}{3} = \frac{3 + 2}{3} = \frac{5}{3}$$

The improper fraction $\frac{5}{3}$ is equivalent to the mixed number $1\frac{2}{3}$.

Example: Convert the improper fraction $\frac{9}{2}$ into a mixed number.

- We wish to divide the numerator by the denominator: $9 \div 2$.
- What is the largest multiple of 2 that isn't bigger than 9? The answer is 8.
- Divide 8 by the denominator: $8 \div 2 = 4$. The whole number is 4.
- Subtract the answer to the second step from the numerator: $9 - 8 = 1$. The remainder is 1. Put the remainder over the original denominator to get $\frac{1}{2}$.
- The answer to $9 \div 2$ is 4 with a remainder of 1, which is the same as $4\frac{1}{2}$.

The mixed number $4\frac{1}{2}$ is equivalent to the improper fraction $\frac{9}{2}$.

Example: Convert the improper fraction $\frac{23}{7}$ into a mixed number.

- We wish to divide the numerator by the denominator: $23 \div 7$.
- What is the largest multiple of 7 that isn't bigger than 23? The answer is 21.
- Divide 21 by the denominator: $21 \div 7 = 3$. The whole number is 3.
- Subtract the answer to the second step from the numerator: $23 - 21 = 2$. The remainder is 2. Put the remainder over the original denominator to get $\frac{2}{7}$.
- The answer to $23 \div 7$ is 3 with a remainder of 2, which is the same as $3\frac{2}{7}$.

The mixed number $3\frac{2}{7}$ is equivalent to the improper fraction $\frac{23}{7}$.

Example: Convert the improper fraction $\frac{43}{5}$ into a mixed number.

- We wish to divide the numerator by the denominator: $43 \div 5$.
- What is the largest multiple of 5 that isn't bigger than 43? The answer is 40.
- Divide 40 by the denominator: $40 \div 5 = 8$. The whole number is 8.
- Subtract the answer to the second step from the numerator: $43 - 40 = 3$. The remainder is 3. Put the remainder over the original denominator to get $\frac{3}{5}$.
- The answer to $43 \div 5$ is 8 with a remainder of 3, which is the same as $8\frac{3}{5}$.

The mixed number $8\frac{3}{5}$ is equivalent to the improper fraction $\frac{43}{5}$.

Note: If you're not already familiar with remainders, it may help to review division with remainders before proceeding.

Date: _____   Score: ___/8   Name: _____

Chapter 16 Exercises – Part A

Directions: Convert each mixed number into an improper fraction.

❶

$$4\frac{2}{3} =$$

❷

$$8\frac{1}{2} =$$

❸

$$3\frac{4}{5} =$$

❹

$$1\frac{3}{4} =$$

❺

$$2\frac{5}{9} =$$

❻

$$6\frac{3}{7} =$$

❼

$$9\frac{1}{6} =$$

❽

$$5\frac{7}{8} =$$

❖ Check your answers using the Answer Key at the back of the book.

Date: _____   Score: ___/8   Name: _____

Chapter 16 Exercises – Part B

Directions: Convert each improper fraction into a mixed number.

❾

$$\frac{8}{3} =$$

❿

$$\frac{22}{5} =$$

⓫

$$\frac{29}{6} =$$

⓬

$$\frac{11}{2} =$$

⓭

$$\frac{31}{9} =$$

⓮

$$\frac{27}{4} =$$

⓯

$$\frac{9}{7} =$$

⓰

$$\frac{53}{8} =$$

❖Check your answers using the Answer Key at the back of the book.

17 RECIPROCALS OF MIXED NUMBERS

Find the reciprocal of a mixed number in two steps:
- First convert the mixed number to an improper fraction using the method from Chapter 16: Multiply the whole number with the denominator, and then add the original numerator to make the new numerator. Write this number over the original denominator. See the example below.

$$5\frac{3}{4} = \frac{5 \times 4 + 3}{4} = \frac{20 + 3}{4} = \frac{23}{4}$$

- Swap the numerator and denominator of the improper fraction like we did in Chapter 8 (which introduced reciprocals). For example, the reciprocal of $\frac{23}{4}$ is $\frac{4}{23}$. Since $5\frac{3}{4} = \frac{23}{4}$, the reciprocal of $5\frac{3}{4}$ is also $\frac{4}{23}$.

Definitions

- **numerator**: the number that appears in the top of a fraction.
- **denominator**: the number that appears in the bottom of a fraction.
- **reciprocal**: what you get when you swap the numerator and denominator.
- **mixed number**: a whole number and fraction combined together, such as $5\frac{3}{4}$.
- **improper fraction**: a fraction for which the numerator is larger than the denominator, such as $\frac{23}{4}$.

Example: Find the reciprocal of $4\frac{1}{3}$.
- First convert the mixed number into an improper fraction using the method from Chapter 16:

$$4\frac{1}{3} = \frac{4 \times 3 + 1}{3} = \frac{12 + 1}{3} = \frac{13}{3}$$

- Swap the numerator and denominator of the improper fraction to get $\frac{3}{13}$.

The reciprocal of $4\frac{1}{3}$ is $\frac{3}{13}$.

Example: Find the reciprocal of $6\frac{2}{3}$.

- First convert the mixed number into an improper fraction using the method from Chapter 16:

$$6\frac{2}{3} = \frac{6 \times 3 + 2}{3} = \frac{18 + 2}{3} = \frac{20}{3}$$

- Swap the numerator and denominator of the improper fraction to get $\frac{3}{20}$.

The reciprocal of $6\frac{2}{3}$ is $\frac{3}{20}$.

Example: Find the reciprocal of $1\frac{8}{9}$.

- First convert the mixed number into an improper fraction using the method from Chapter 16:

$$1\frac{8}{9} = \frac{1 \times 9 + 8}{9} = \frac{9 + 8}{9} = \frac{17}{9}$$

- Swap the numerator and denominator of the improper fraction to get $\frac{9}{17}$.

The reciprocal of $1\frac{8}{9}$ is $\frac{9}{17}$.

Example: Find the reciprocal of $3\frac{4}{7}$.

- First convert the mixed number into an improper fraction using the method from Chapter 16:

$$3\frac{4}{7} = \frac{3 \times 7 + 4}{7} = \frac{21 + 4}{7} = \frac{25}{7}$$

- Swap the numerator and denominator of the improper fraction to get $\frac{7}{25}$.

The reciprocal of $3\frac{4}{7}$ is $\frac{7}{25}$.

Example: Find the reciprocal of $9\frac{1}{4}$.

- First convert the mixed number into an improper fraction using the method from Chapter 16:

$$9\frac{1}{4} = \frac{9 \times 4 + 1}{4} = \frac{36 + 1}{4} = \frac{37}{4}$$

- Swap the numerator and denominator of the improper fraction to get $\frac{4}{37}$.

The reciprocal of $9\frac{1}{4}$ is $\frac{4}{37}$.

Date: _____   Score: ___/8   Name: _____

Chapter 17 Exercises – Part A

Directions: Find the reciprocal of each mixed number.

❶

$3\dfrac{1}{2}$

❷

$4\dfrac{5}{8}$

❸

$1\dfrac{4}{5}$

❹

$7\dfrac{3}{4}$

❺

$5\dfrac{4}{5}$

❻

$2\dfrac{4}{7}$

❼

$8\dfrac{2}{5}$

❽

$9\dfrac{8}{9}$

❖ Check your answers using the Answer Key at the back of the book.

Date: _____   Score: ___/8   Name: _____

Chapter 17 Exercises – Part B

Directions: Find the reciprocal of each mixed number.

❾

$$5\frac{7}{8}$$

❿

$$6\frac{1}{9}$$

⓫

$$2\frac{2}{9}$$

⓬

$$1\frac{1}{2}$$

⓭

$$9\frac{2}{3}$$

⓮

$$4\frac{5}{9}$$

⓯

$$3\frac{1}{3}$$

⓰

$$8\frac{3}{7}$$

❖ Check your answers using the Answer Key at the back of the book.

18 ARITHMETIC WITH MIXED NUMBERS

One way to add, subtract, multiply, or divide two mixed numbers is:

- First convert the mixed number to an improper fraction using the method from Chapter 16: Multiply the whole number with the denominator, and then add the original numerator to make the new numerator. Write this number over the original denominator. See the example below.

$$5\frac{3}{4} = \frac{5 \times 4 + 3}{4} = \frac{20 + 3}{4} = \frac{23}{4}$$

- Add, subtract, multiply, or divide the improper fractions using the methods from Chapters 5, 6, 7, or 9. To add or subtract fractions, first find a common denominator. To divide two fractions, multiply the first fraction with the reciprocal of the second fraction. See the examples on the following pages.

- If the answer to the previous step is an improper fraction, convert it to a mixed number using the method from Chapter 16: Divide the numerator by the denominator to get the whole number and fraction (where the fraction equals the remainder over the original denominator). See the example below.

$$\frac{38}{5} = 38 \div 5 = 7 \text{ R}3 = 7 + \frac{3}{5} = 7\frac{3}{5}$$

Definitions

- **numerator**: the number that appears in the top of a fraction.
- **denominator**: the number that appears in the bottom of a fraction.
- **common denominator**: when two fractions each have the same denominator.
- **reciprocal**: what you get when you swap the numerator and denominator.
- **mixed number**: a whole number and fraction combined together, such as $5\frac{3}{8}$.
- **improper fraction**: a fraction for which the numerator is larger than the denominator, such as $\frac{43}{8}$.

Example: Add the mixed numbers $5\frac{2}{3}$ and $4\frac{1}{2}$.

- First convert the mixed numbers into improper fractions using the method from Chapter 16:

$$5\frac{2}{3} = \frac{5 \times 3 + 2}{3} = \frac{15 + 2}{3} = \frac{17}{3}$$
$$4\frac{1}{2} = \frac{4 \times 2 + 1}{2} = \frac{8 + 1}{2} = \frac{9}{2}$$

- Next find a common denominator using the method from Chapter 5. The least common multiple of 3 and 2 is the number 6.

$$\frac{17}{3} + \frac{9}{2} = \frac{17 \times 2}{3 \times 2} + \frac{9 \times 3}{2 \times 3} = \frac{34}{6} + \frac{27}{6} = \frac{34 + 27}{6} = \frac{61}{6}$$

- Convert the improper fraction $\frac{61}{6}$ into a mixed number using the method from Chapter 16. Note that $60 \div 10 = 6$ and $61 - 60 = 1$ such that $61 \div 6 = 10$ R1.

$$\frac{61}{6} = 10\frac{1}{6}$$

Our final answer is $10\frac{1}{6}$.

Example: Perform the subtraction $4\frac{2}{9} - 1\frac{5}{6}$.

- First convert the mixed numbers into improper fractions using the method from Chapter 16:

$$4\frac{2}{9} = \frac{4 \times 9 + 2}{9} = \frac{36 + 2}{9} = \frac{38}{9}$$
$$1\frac{5}{6} = \frac{1 \times 6 + 5}{6} = \frac{6 + 5}{6} = \frac{11}{6}$$

- Next find a common denominator using the method from Chapter 6. The least common multiple of 9 and 6 is the number 18.

$$\frac{38}{9} - \frac{11}{6} = \frac{38 \times 2}{9 \times 2} - \frac{11 \times 3}{6 \times 3} = \frac{76}{18} - \frac{33}{18} = \frac{76 - 33}{18} = \frac{43}{18}$$

- Convert the improper fraction $\frac{43}{18}$ into a mixed number using the method from Chapter 16. Note that $36 \div 18 = 2$ and $43 - 36 = 7$ such that $36 \div 18 = 2$ R7.

$$\frac{43}{18} = 2\frac{7}{18}$$

Our final answer is $2\frac{7}{18}$.

Example: Multiply the fractions $2\frac{1}{4}$ and $3\frac{2}{3}$.

- First convert the mixed numbers into improper fractions using the method from Chapter 16:

$$2\frac{1}{4} = \frac{2 \times 4 + 1}{4} = \frac{8 + 1}{4} = \frac{9}{4}$$
$$3\frac{2}{3} = \frac{3 \times 3 + 2}{3} = \frac{9 + 2}{3} = \frac{11}{3}$$

- Next multiply $\frac{9}{4}$ and $\frac{11}{3}$ using the method from Chapter 7.

$$\frac{9}{4} \times \frac{11}{3} = \frac{9 \times 11}{4 \times 3} = \frac{99}{12}$$

- The fraction $\frac{99}{12}$ can be reduced since 99 and 12 are both divisible by 3. Divide the numerator and denominator both by 3.

$$\frac{99}{12} = \frac{99 \div 3}{12 \div 3} = \frac{33}{4}$$

- Convert the improper fraction $\frac{33}{4}$ into a mixed number using the method from Chapter 16. Note that $32 \div 4 = 8$ and $33 - 32 = 1$ such that $33 \div 4 = 8$ R1.

$$\frac{33}{4} = 8\frac{1}{4}$$

Our final answer is $8\frac{1}{4}$.

Calculator check: Enter $\left(2 + \frac{1}{4}\right) \times \left(3 + \frac{2}{3}\right)$ on your calculator. Be sure to use parentheses. (If your calculator doesn't have parentheses, first find $2 + \frac{1}{4} = 2.25$ and $3 + \frac{2}{3} = 3.66666667$ separately and then multiply 2.25×3.66666667.) Compare your answer with $8 + \frac{1}{4}$. Both equal 8.25 (though on a calculator you could be slightly off due to rounding).

Example: Perform the division $5\frac{1}{2} \div 3\frac{3}{4}$.

- First convert the mixed numbers into improper fractions using the method from Chapter 16:

$$5\frac{1}{2} = \frac{5 \times 2 + 1}{2} = \frac{10 + 1}{2} = \frac{11}{2}$$
$$3\frac{3}{4} = \frac{3 \times 4 + 3}{4} = \frac{12 + 3}{4} = \frac{15}{4}$$

- Next perform the division $\frac{11}{2} \div \frac{15}{4}$ using the method from Chapter 9. Recall that dividing by a fraction is equivalent to multiplying by its reciprocal. The reciprocal of $\frac{15}{4}$ is $\frac{4}{15}$.

$$\frac{11}{2} \div \frac{15}{4} = \frac{11}{2} \times \frac{4}{15} = \frac{11 \times 4}{2 \times 15} = \frac{44}{30}$$

- The fraction $\frac{44}{30}$ can be reduced since 44 and 30 are both divisible by 2. Divide the numerator and denominator both by 2.

$$\frac{44}{30} = \frac{44 \div 2}{30 \div 2} = \frac{22}{15}$$

- Convert the improper fraction $\frac{22}{15}$ into a mixed number using the method from Chapter 16. Note that $15 \div 15 = 1$ and $22 - 15 = 7$ such that $22 \div 15 = 1$ R7.

$$\frac{22}{15} = 1\frac{7}{15}$$

Our final answer is $1\frac{7}{15}$.

Calculator check: Enter $\left(5 + \frac{1}{2}\right) \div \left(3 + \frac{3}{4}\right)$ on your calculator. Be sure to use parentheses. (If your calculator doesn't have parentheses, first find $5 + \frac{1}{2} = 5.5$ and $3 + \frac{3}{4} = 3.75$ separately and then divide $5.5 \div 3.75$.) Compare your answer with $1 + \frac{7}{15}$. Both equal $1.4\overline{6}$.

Date: _____   Score: ___/5   Name: _____

Chapter 18 Exercises – Part A

Directions: Add the mixed numbers by first finding a common denominator.

❶

$$1\frac{3}{5} + 2\frac{1}{3} =$$

❷

$$3\frac{1}{2} + 4\frac{2}{3} =$$

❸

$$5\frac{3}{4} + 2\frac{1}{6} =$$

❹

$$7\frac{1}{6} + 8\frac{5}{6} =$$

❺

$$6\frac{4}{5} + 4\frac{1}{2} =$$

❖ Check your answers using the Answer Key at the back of the book.

Date: _____   Score: ___/5   Name: _____

Chapter 18 Exercises – Part B

Directions: Subtract the mixed numbers by first finding a common denominator.

❻

$$5\frac{1}{3} - 2\frac{1}{2} =$$

❼

$$6\frac{1}{4} - 3\frac{2}{5} =$$

❽

$$4\frac{5}{8} - 1\frac{1}{6} =$$

❾

$$8 - 4\frac{5}{6} =$$

❿

$$9\frac{5}{8} - 7 =$$

❖ Check your answers using the Answer Key at the back of the book.

Date: _____   Score: ___/5   Name: _____

Chapter 18 Exercises – Part C

Directions: Multiply the given mixed numbers.

⑪

$$4\frac{1}{2} \times 2\frac{2}{3} =$$

⑫

$$6\frac{2}{3} \times 4\frac{1}{5} =$$

⑬

$$8\frac{3}{4} \times 3\frac{1}{3} =$$

⑭

$$6\frac{2}{3} \times 5 =$$

⑮

$$2\frac{1}{2} \times 1\frac{3}{4} =$$

❖ Check your answers using the Answer Key at the back of the book.

Date: _____ Score: ___/5 Name: _____

Chapter 18 Exercises – Part D

Directions: Divide the given mixed numbers.

16

$$6\frac{2}{3} \div 3\frac{1}{5} =$$

17

$$7\frac{4}{9} \div 4\frac{1}{6} =$$

18

$$3\frac{3}{4} \div \frac{1}{2} =$$

19

$$6 \div 2\frac{5}{8} =$$

20

$$8\frac{4}{9} \div 3 =$$

❖ Check your answers using the Answer Key at the back of the book.

19 RATIOS AND PROPORTIONS

A ratio expresses a fixed relationship as a fraction. For example, if a town of 200 people has 300 cars, we can say that the ratio of people to cars in that town is 2 to 3. We sometimes express a ratio using a colon (:), as in the ratio is 2:3, but we could equivalently express it as a fraction, as in $\frac{2}{3}$.

A proportion expresses an equality between two ratios for quantities that obey a direct proportion. For example, suppose that a grocery store sells hot dogs in packages of 10 and hot dog buns in packages of 8. If you buy one package of each, you will have 10 hot dogs for every 8 hot dog buns, meaning that the ratio of hot dogs to hot dog buns is 10:8. We can set up a proportion, saying that the ratio 10:8 equals 5:4. This is sometimes written in the form 10:8::5:4, which reads "10 is to 8 as 5 is to 4." We could express this same information using fractions, like this:

$$\frac{10}{8} = \frac{5}{4}$$

To see that the above proportion holds, multiply the numerator and denominator of $\frac{5}{4}$ by 2. You will get $\frac{10}{8}$ since $5 \times 2 = 10$ and $4 \times 2 = 8$, which shows that $\frac{10}{8}$ equals $\frac{5}{4}$.

Sometimes when we have a proportion, we wish to solve for an unknown quantity. For example, in the proportion shown below, the box ($\square$) represents an unknown.

$$\frac{1}{2} = \frac{\square}{8}$$

In this example, the box ($\square$) is 4 since 4 is one-half of 8. One way to solve a proportion problem is to make equivalent fractions. This is illustrated in the following examples.

Definitions

- **ratio**: a fixed relationship that is expressed as a fraction.
- **proportion**: an equality between two ratios. For example, the ratio 6:3 is equal to the ratio 2:1 (in both cases, the numerator is twice the denominator).

Example: Write a number in the box that makes the following proportion true.

$$\frac{6}{9} = \frac{\square}{27}$$

- Multiply the numerator and denominator of $\frac{6}{9}$ by 3 since $9 \times 3 = 27$.

$$\frac{6}{9} = \frac{6 \times 3}{9 \times 3} = \frac{18}{27}$$

- Since $\frac{18}{27}$ is equivalent to $\frac{6}{9}$, the new numerator must be 18.

$$\frac{6}{9} = \frac{\boxed{18}}{27}$$

Example: Write a number in the box that makes the following proportion true.

$$\frac{4}{5} = \frac{8}{\square}$$

- Multiply the numerator and denominator of $\frac{4}{5}$ by 2 since $4 \times 2 = 8$.

$$\frac{4}{5} = \frac{4 \times 2}{5 \times 2} = \frac{8}{10}$$

- Since $\frac{8}{10}$ is equivalent to $\frac{4}{5}$, the new denominator must be 10.

$$\frac{4}{5} = \frac{8}{\boxed{10}}$$

Example: Write a number in the box that makes the following proportion true.

$$\frac{16}{12} = \frac{\square}{18}$$

- Since 12 doesn't divide evenly into 18, we'll first reduce $\frac{16}{12}$.

$$\frac{16}{12} = \frac{16 \div 4}{12 \div 4} = \frac{4}{3} \quad \text{which means that} \quad \frac{4}{3} = \frac{\square}{18}$$

- Now multiply the numerator and denominator of $\frac{4}{3}$ by 6 since $3 \times 6 = 18$.

$$\frac{4}{3} = \frac{4 \times 6}{3 \times 6} = \frac{24}{18}$$

- Since $\frac{4}{3}$ is equivalent to both $\frac{16}{12}$ and $\frac{24}{18}$, the new numerator must be 24.

$$\frac{16}{12} = \frac{\boxed{24}}{18}$$

Date: _____   Score: ___/8   Name: _____

Chapter 19 Exercises – Part A

Directions: Write a number in each box that makes the proportion true.

❶

$$\frac{12}{8} = \frac{\boxed{}}{24}$$

❺

$$\frac{4}{7} = \frac{\boxed{}}{28}$$

❷

$$\frac{\boxed{}}{15} = \frac{3}{5}$$

❻

$$\frac{\boxed{}}{4} = \frac{9}{3}$$

❸

$$\frac{3}{6} = \frac{\boxed{}}{10}$$

❼

$$\frac{28}{44} = \frac{\boxed{}}{11}$$

❹

$$\frac{\boxed{}}{3} = \frac{24}{18}$$

❽

$$\frac{\boxed{}}{4} = \frac{15}{10}$$

❖ Check your answers using the Answer Key at the back of the book.

Date: _____ Score: ___/8 Name: _____

Chapter 19 Exercises – Part B

Directions: Write a number in each box that makes the proportion true.

❾

$$\frac{9}{8} = \frac{\boxed{}}{72}$$

❿

$$\frac{\boxed{}}{20} = \frac{45}{25}$$

⓫

$$\frac{3}{2} = \frac{\boxed{}}{16}$$

⓬

$$\frac{\boxed{}}{8} = \frac{35}{28}$$

⓭

$$\frac{5}{100} = \frac{\boxed{}}{20}$$

⓮

$$\frac{\boxed{}}{28} = \frac{24}{21}$$

⓯

$$\frac{13}{13} = \frac{\boxed{}}{95}$$

⓰

$$\frac{\boxed{}}{40} = \frac{6}{16}$$

❖ Check your answers using the Answer Key at the back of the book.

20 LONG DIVISION

Chapters 10 and 11 presented one method for converting a fraction into a decimal. In this chapter, we will explore a different method. Any fraction can be converted into a decimal using the method of long division.

A fraction consists of a numerator divided by a denominator.

$$\frac{\text{numerator}}{\text{denominator}} = \text{numerator} \div \text{denominator} = \text{denominator} \overline{)\text{numerator}}$$

For example, consider the fraction $\frac{3}{4}$. This means 3 divided by 4.

$$\frac{3}{4} = 3 \div 4 = 4\overline{)3}$$

When we divide two numbers, like $28 \div 7$, the number being dividing into (28) is called the dividend, the number we're dividing by (7) is called the divisor, and the result of the division (in this example, it's 4) is called the quotient.

$$\text{dividend} \div \text{divisor} = \text{quotient} \qquad \text{divisor}\overline{)\text{dividend}}^{\text{quotient}}$$

$$28 \div 7 = 4 \qquad 7\overline{)28}^{\,4}$$

Definitions

- **quotient**: the answer to a division problem.
- **divisor**: the number we're dividing by.
- **dividend**: the number being divided into.

Example: Convert the fraction $\frac{7}{8}$ into a decimal.

Set up a long division problem with the numerator (7) as the dividend and the denominator (8) as the divisor.

$$8\,)\overline{\,7\,}$$

Since 8 is larger than 7, we will add a point zero to 7.

$$8\,)\overline{\,7.0\,}$$

We'd like to multiply 8 by something to make a number close to 7.0 (but not bigger than 7.0). We can make $8 \times 0.8 = 6.4$. Write the 0.8 at the top. Subtract 6.4 from 7.0. Add a zero to the difference (0.6) to make 0.60. See below.

$$
\begin{array}{r}
0.8 \\
8\,)\overline{7.0} \\
-6.4 \\
\hline
0.60
\end{array}
$$

Now we'd like to multiply 8 by something to make a number close to 0.60 (but not bigger than 0.60). We can make $8 \times 0.07 = 0.56$. Join the 0.07 to the top. Subtract 0.56 from 0.60. Add a zero to the difference (0.04) to make 0.040. See below.

$$
\begin{array}{r}
0.87 \\
8\,)\overline{7.0} \\
-6.4 \\
\hline
0.60 \\
-0.56 \\
\hline
0.040
\end{array}
$$

Next we're dividing 0.040 by 8. Note that $8 \times 0.005 = 0.040$. Join the 0.005 to the top. Subtract 0.040 from 0.040. Since the difference is zero, we're finished.

$$
\begin{array}{r}
0.875 \\
8\,)\overline{7.0} \\
-6.4 \\
\hline
0.60 \\
-0.56 \\
\hline
0.040 \\
-0.040 \\
\hline
0.000
\end{array}
$$

The fraction $\frac{7}{8}$ is equivalent to the decimal 0.875.

Example: Convert the fraction $\frac{3}{11}$ into a decimal.

Set up a long division problem with the numerator (3) as the dividend and the denominator (11) as the divisor.

$$11\overline{)\,3.0}$$

We'd like to multiply 11 by something to make a number close to 3.0 (but not bigger than 3.0). We can make $11 \times 0.2 = 2.2$. Write the 0.2 at the top. Subtract 2.2 from 3.0. Add a zero to the difference (0.8) to make 0.80. See below.

$$
\begin{array}{r}
0.2 \\
11\overline{)\,3.0} \\
-2.2 \\
\hline
0.80
\end{array}
$$

Now we'd like to multiply 11 by something to make a number close to 0.80 (but not bigger than 0.80). We can make $11 \times 0.07 = 0.77$. Join the 0.07 to the top. Subtract 0.77 from 0.80. Add a zero to the difference (0.03) to make 0.030. See below.

$$
\begin{array}{r}
0.27 \\
11\overline{)\,3.0} \\
-2.2 \\
\hline
0.80 \\
-0.77 \\
\hline
0.030
\end{array}
$$

Next we're dividing 0.030 by 11. Note that $11 \times 0.002 = 0.022$. Join the 0.002 to the top. Subtract 0.022 from 0.030. We get 0.008 as the difference. Note that $11 \times 0.0007 = 0.0077$. When we join this 0.0007 to the top, we see that the 27 is repeating to make 0.2727. This is a repeating decimal. The pattern will go on forever.

$$
\begin{array}{r}
0.2727 \\
11\overline{)\,3.0} \\
-2.2 \\
\hline
0.80 \\
-0.77 \\
\hline
0.030 \\
-0.022 \\
\hline
0.0080
\end{array}
$$

The fraction $\frac{3}{11}$ is equivalent to the repeating decimal $0.\overline{27}$ (recall Chapter 11).

Example: Below are some more examples. Study them and try to think your way through the math. If necessary, compare these to the two previous examples.

Convert $\frac{3}{4}$ into a decimal. Convert $\frac{9}{2}$ into a decimal. Convert $\frac{2}{3}$ into a decimal.

$$
\begin{array}{r}
0.75 \\
4\overline{)3.0} \\
-2.8 \\
\hline
0.20 \\
-0.20 \\
\hline
0.00
\end{array}
$$
$4 \times 0.7 = 2.8$

$4 \times 0.05 = 0.20$

$$\frac{3}{4} = 0.75$$

$$
\begin{array}{r}
4.5 \\
2\overline{)9} \\
-8 \\
\hline
1.0 \\
-1.0 \\
\hline
0.0
\end{array}
$$
$2 \times 4 = 8$

$2 \times 0.5 = 1.0$

$$\frac{9}{2} = 4.5$$

$$
\begin{array}{r}
0.66 \\
3\overline{)2.0} \\
-1.8 \\
\hline
0.20 \\
-0.18 \\
\hline
0.020
\end{array}
$$
$3 \times 0.6 = 1.8$

$3 \times 0.06 = 0.18$

$$\frac{2}{3} = 0.\bar{6} \text{ (note the bar)}$$

Convert $\frac{5}{16}$ into a decimal.   Convert $\frac{16}{33}$ into a decimal.

$$
\begin{array}{r}
0.3125 \\
16\overline{)5.0} \\
-4.8 \\
\hline
0.20 \\
-0.16 \\
\hline
0.040 \\
-0.032 \\
\hline
0.008 \\
-0.008 \\
\hline
0.000
\end{array}
$$
$16 \times 0.3 = 4.8$

$16 \times 0.01 = 0.16$

$16 \times 0.002 = 0.032$

$16 \times 0.0005 = 0.0080$

$$\frac{5}{16} = 0.3125$$

$$
\begin{array}{r}
0.484848 \\
33\overline{)16.0} \\
-13.2 \\
\hline
2.80 \\
-2.64 \\
\hline
0.160 \\
-0.132 \\
\hline
0.0280 \\
-0.0264 \\
\hline
0.00160
\end{array}
$$
$33 \times 0.4 = 13.2$

$33 \times 0.08 = 2.64$

$33 \times 0.004 = 0.132$

$33 \times 0.0008 = 0.0264$

$$\frac{16}{33} = 0.\overline{48} \text{ (note the bar)}$$

Date: _____   Score: ___/2   Name: _____

Chapter 20 Exercises – Part A

Directions: Convert each fraction to a decimal using long division.

❶

$$\frac{3}{8} =$$

❷

$$\frac{5}{11} =$$

❖ Check your answers using the Answer Key at the back of the book.

Date: _____   Score: ___/2   Name: _____

Chapter 20 Exercises – Part B

Directions: Convert each fraction to a decimal using long division.

❸

$$\frac{7}{6} =$$

❹

$$\frac{4}{9} =$$

❖ Check your answers using the Answer Key at the back of the book.

Date: _____   Score: ___/2   Name: _____

Chapter 20 Exercises – Part C

Directions: Convert each fraction to a decimal using long division.

❺

$$\frac{7}{16} =$$

❻

$$\frac{16}{75} =$$

❖ Check your answers using the Answer Key at the back of the book.

Date: _____   Score: ___/2   Name: _____

Chapter 20 Exercises – Part D

Directions: Convert each fraction to a decimal using long division.

❼

$$\frac{17}{40} =$$

❽

$$\frac{5}{12} =$$

❖ Check your answers using the Answer Key at the back of the book.

Date: _____   Score: ___/20   Name: _____

POSTTEST ASSESSMENT

The purpose of this posttest is to help you (or a teacher or parent) see how much you have learned about fractions. If you took the pretest assessment before you started using this workbook, compare your score on the posttest to your score on the pretest to see how much you have improved.

When you finish the posttest, check your answers with the key on page 184. Then check the explanations offered on page 245. If you struggle with any of the problems on this posttest, it would be wise to find the chapter in this book that has similar problems and review the material.

❶ Express $\frac{15}{40}$ as a reduced fraction.

❷ Which is bigger, $\frac{2}{3}$ or $\frac{5}{8}$?

❖ Continued on the next page.

❸ What is the reciprocal of $\frac{4}{7}$?

❹ What is the reciprocal of 5?

❺ Evaluate $\frac{3}{4} + \frac{1}{6}$. Express your answer as a reduced fraction.

❻ Evaluate $\frac{3}{5} - \frac{2}{9}$. Express your answer as a reduced fraction.

❼ Evaluate $\frac{2}{3} \times \frac{5}{4}$. Express your answer as a reduced fraction.

❖ Continued on the next page.

❽ Evaluate $\frac{5}{12} \div \frac{3}{4}$. Express your answer as a reduced fraction.

❾ Express 0.75 as a reduced fraction.

❿ Express $\frac{5}{8}$ in decimal form.

⓫ Express 0.4 as a percentage.

⓬ Express $\frac{5}{2}$ as a percentage.

❖ Continued on the next page.

⓭ Express 35% as a reduced fraction.

⓮ Express 150% in decimal form.

⓯ Express $\frac{8}{3}$ as a repeating decimal.

⓰ Express the repeating decimal $0.\overline{45}$ as a reduced fraction.

❖ Continued on the next page.

⓱ Express the improper fraction $\frac{20}{3}$ as a mixed number.

⓲ Express the mixed number $2\frac{1}{4}$ as an improper fraction.

⓳ Evaluate $3\frac{2}{5} + 5\frac{3}{4}$. Express your answer as a mixed number.

⓴ A parking lot has 6 red cars and 12 blue cars. What fraction of the cars are blue?

❖ Check your answers on the next page.

Check your answers to the posttest.

❶ $\frac{3}{8}$ ❷ $\frac{2}{3}$ ❸ $\frac{7}{4}$ ❹ $\frac{1}{5}$ ❺ $\frac{11}{12}$

❻ $\frac{17}{45}$ ❼ $\frac{5}{6}$ ❽ $\frac{5}{9}$ ❾ $\frac{3}{4}$ ❿ 0.625

⓫ 40%   ⓬ 250%   ⓭ $\frac{7}{20}$ ⓮ 1.5   ⓯ $2.\overline{6}$

⓰ $\frac{5}{11}$ ⓱ $6\frac{2}{3}$ ⓲ $\frac{9}{4}$ ⓳ $9\frac{3}{20}$ ⓴ $\frac{2}{3}$

Explanations: You can find the explanation to every problem of the posttest on page 245.

ANSWER KEY

Pretest Assessment

❶ $\frac{3}{8}$

❷ $\frac{2}{3}$

❸ $\frac{7}{4}$

❹ $\frac{1}{5}$

❺ $\frac{11}{12}$

❻ $\frac{17}{45}$

❼ $\frac{5}{6}$ (If you get $\frac{10}{12}$, you didn't reduce your answer. See Chapter 2.)

❽ $\frac{5}{9}$ (If you get $\frac{20}{36}$ or $\frac{10}{18}$, you didn't fully reduce your answer. See Chapter 2.)

❾ $\frac{3}{4}$ (If you get $\frac{75}{100}$, you didn't reduce your answer. See Chapter 2.)

❿ 0.625

⓫ 40%

⓬ 250%

⓭ $\frac{7}{20}$ (If you get $\frac{35}{100}$, you didn't reduce your answer. See Chapter 2.)

⓮ 1.5 (If your answer is 1.50, it's customary to omit the trailing zero.)

⓯ $2.\overline{6}$ (This means 2.666666... where the 6 repeats forever. See Chapter 11.)

⑯ $\frac{5}{11}$ (If you get $\frac{45}{99}$ or $\frac{15}{33}$, you didn't reduce your answer. Note that $\frac{9}{20}$ and $\frac{45}{100}$ are both incorrect: $0.\overline{45}$ means 0.45454545 with the 45 repeating forever. Note that $0.\overline{45}$ doesn't mean the same thing as 0.45. See Chapter 11.)

⑰ $6\frac{2}{3}$ **⑱** $\frac{9}{4}$

⑲ $9\frac{3}{20}$ (If you get $\frac{183}{20}$, you forgot to convert the improper fraction to a mixed number, as shown in Chapter 16. If you get $8\frac{23}{20}$, your answer is correct, but your answer isn't in standard form. Note that $8\frac{23}{20} = 9\frac{3}{20}$ because $\frac{23}{20} = 1\frac{3}{20}$.)

⑳ $\frac{2}{3}$ (If you get $\frac{12}{18}, \frac{6}{9}$, or $\frac{4}{6}$, you didn't fully reduce your answer. See Chapter 2.)

Chapter 1 – Pie Slices

Part A

❶ $\frac{1}{2}$ **❷** $\frac{1}{3}$ **❸** $\frac{3}{4}$

❹ $\frac{1}{5}$ **❺** $\frac{4}{5}$ **❻** $\frac{1}{6}$

❼ $\frac{4}{7}$ **❽** $\frac{7}{8}$ **❾** $\frac{4}{9}$

Part B

❿ $\frac{2}{3}$ **⓫** $\frac{1}{2}$ **⓬** $\frac{2}{5}$

⓭ $\frac{5}{7}$ **⓮** $\frac{3}{8}$ **⓯** $\frac{8}{9}$

⓰ $\frac{1}{8}$ **⓱** $\frac{2}{7}$ **⓲** $\frac{5}{9}$

Chapter 2 – Reducing Fractions

Part A

❶ $\frac{2}{3}$ (Divide the numerator and denominator by 3.)

❷ $\frac{3}{4}$ (Divide the numerator and denominator by 3.
If your answer is $\frac{1}{3}$, check the problem number carefully.)

❸ $\frac{1}{2}$ (Divide the numerator and denominator by 7.
If your answer is $\frac{3}{4}$, check the problem number carefully.)

❹ 9 (Note that $\frac{9}{1} = 9 \div 1 = 9$.
If your answer is 2, check the problem number carefully.)

❺ $\frac{1}{3}$ (Divide the numerator and denominator by 2.
If your answer is $\frac{1}{2}$, check the problem number carefully.)

❻ 2 (Divide the numerator and denominator by 2.
Note that $\frac{4}{2} = 4 \div 2 = 2$.
If your answer is $\frac{3}{5}$, check the problem number carefully.)

❼ $\frac{3}{5}$ (Divide the numerator and denominator by 5.
If your answer is 9, check the problem number carefully.)

❽ $\frac{5}{6}$ (Divide the numerator and denominator by 2.)

Part B

❾ $\frac{2}{5}$ (Divide the numerator and denominator by 4.

If your answer is $\frac{4}{10}$, you need to reduce it again.)

❿ 4 (Divide the numerator and denominator by 3. Note that $\frac{12}{3} =$

$12 \div 3 = 4$. If your answer is $\frac{5}{2}$, check the problem number carefully.)

⓫ $\frac{2}{3}$ (Divide the numerator and denominator by 12.

If your answer is $\frac{4}{6}, \frac{6}{9}, \frac{8}{12}$, or $\frac{12}{18}$, you need to reduce it again.

If your answer is 4, check the problem number carefully.)

⓬ $\frac{7}{9}$ (Divide the numerator and denominator by 4. If your answer is $\frac{14}{18}$,

you need to reduce it again. If your answer is $\frac{4}{5}$, you made an

arithmetic mistake or you misread the 36 as 35.

If your answer is 6, check the problem number carefully.)

⓭ $\frac{5}{2}$ (Divide the numerator and denominator by 2.

If your answer is $\frac{2}{3}$, check the problem number carefully.)

⓮ 6 (Divide the numerator and denominator by 5.

Note that $\frac{30}{5} = 30 \div 5 = 6$.

If your answer is $\frac{5}{4}$, check the problem number carefully.)

⓯ $\frac{5}{4}$ (Divide the numerator and denominator by 4.

If your answer is $\frac{10}{8}$, you need to reduce it again.

If your answer is $\frac{7}{9}$, check the problem number carefully.)

⓰ 1 (Note that $\frac{1}{1} = 1 \div 1 = 1$.)

Chapter 3 – Common Denominators

Part A

❶ $\frac{3}{6}$ and $\frac{4}{6}$ (Multiply the numerator and denominator of $\frac{1}{2}$ by 3, and multiply the numerator and denominator of $\frac{2}{3}$ by 2.)

❷ $\frac{20}{36}$ and $\frac{27}{36}$ (Multiply the numerator and denominator of $\frac{5}{9}$ by 4, and multiply the numerator and denominator of $\frac{3}{4}$ by 9.)

❸ $\frac{3}{24}$ and $\frac{2}{24}$ (Multiply the numerator and denominator of $\frac{1}{8}$ by 3, and multiply the numerator and denominator of $\frac{1}{12}$ by 2.
If your denominators are 48 or 96, your denominator isn't the lowest common denominator: Make 24 instead, or reduce your answers to a denominator of 24.)

❹ $\frac{21}{30}$ and $\frac{8}{30}$ (Multiply the numerator and denominator of $\frac{7}{10}$ by 3, and multiply the numerator and denominator of $\frac{4}{15}$ by 2.
If your denominators are 60, 90, 120, or 150, your denominator isn't the lowest common denominator: Make 30 instead, or reduce your answers to a denominator of 30.)

❺ $\frac{12}{3}$ and $\frac{5}{3}$ (Note that $4 = \frac{4}{1}$. Multiply the numerator and denominator of $\frac{4}{1}$ by 3, but leave $\frac{5}{3}$ the way it is.)

Part B

❻ $\frac{21}{12}$ and $\frac{20}{12}$ (Multiply the numerator and denominator of $\frac{7}{4}$ by 3, and multiply the numerator and denominator of $\frac{5}{3}$ by 4.)

❼ $\frac{44}{48}$ and $\frac{27}{48}$ (Multiply the numerator and denominator of $\frac{11}{12}$ by 4, and multiply the numerator and denominator of $\frac{9}{16}$ by 3.

If your denominators are 96, 144, or 192, your denominator isn't the lowest common denominator: Make 48 instead, or reduce your answers to a denominator of 48.)

❽ $\frac{15}{72}$ and $\frac{14}{72}$ (Multiply the numerator and denominator of $\frac{5}{24}$ by 3, and multiply the numerator and denominator of $\frac{7}{36}$ by 2.

If your denominators are 144 or higher, your denominator isn't the lowest common denominator: Make 72 instead, or reduce your answers to a denominator of 72.)

❾ $\frac{9}{4}$ and $\frac{4}{4}$ (Note that $1 = \frac{1}{1}$. Multiply the numerator and denominator of $\frac{1}{1}$ by 4, but leave $\frac{9}{4}$ the way it is.)

❿ $\frac{9}{18}$ and $\frac{14}{18}$ (Multiply the numerator and denominator of $\frac{1}{2}$ by 9, and multiply the numerator and denominator of $\frac{7}{9}$ by 2.)

Part C

⓫ $\frac{9}{15}$ and $\frac{25}{15}$ (Multiply the numerator and denominator of $\frac{3}{5}$ by 3, and multiply the numerator and denominator of $\frac{5}{3}$ by 5.

⓬ $\frac{36}{100}$ and $\frac{55}{100}$ (Multiply the numerator and denominator of $\frac{9}{25}$ by 4, and multiply the numerator and denominator of $\frac{11}{20}$ by 5.
If your denominators are 200, 300, 400, or 500, your denominator isn't the lowest common denominator: Make 100 instead, or reduce your answers to a denominator of 100.)

⓭ $\frac{10}{2}$ and $\frac{5}{2}$ (Note that $5 = \frac{5}{1}$. Multiply the numerator and denominator of $\frac{5}{1}$ by 2, but leave $\frac{5}{2}$ the way it is.)

⓮ $\frac{24}{45}$ and $\frac{50}{45}$ (Multiply the numerator and denominator of $\frac{8}{15}$ by 3, and multiply the numerator and denominator of $\frac{10}{9}$ by 5.
If your denominators are 90 or 135, your denominator isn't the lowest common denominator: Make 45 instead, or reduce your answers to a denominator of 45.)

⓯ $\frac{3}{8}$ and $\frac{2}{8}$ (Multiply the numerator and denominator of $\frac{3}{8}$ by 1, and multiply the numerator and denominator of $\frac{1}{4}$ by 2.
If your denominators are 16, 24, or 32, your denominator isn't the lowest common denominator: Make 8 instead, or reduce your answers to a denominator of 8.)

Part D

16 $\frac{32}{84}$ and $\frac{15}{84}$ (Multiply the numerator and denominator of $\frac{8}{21}$ by 4, and multiply the numerator and denominator of $\frac{5}{28}$ by 3.

If your denominators are 168, 252, 336, 420, 504, or 588, your denominator isn't the lowest common denominator: Make 84 instead, or reduce your answers to a denominator of 84.)

17 $\frac{81}{72}$ and $\frac{64}{72}$ (Multiply the numerator and denominator of $\frac{9}{8}$ by 9, and multiply the numerator and denominator of $\frac{8}{9}$ by 8.)

18 $\frac{33}{22}$ and $\frac{12}{22}$ (Multiply the numerator and denominator of $\frac{3}{2}$ by 11, and multiply the numerator and denominator of $\frac{6}{11}$ by 2.)

19 $\frac{18}{3}$ and $\frac{8}{3}$ (Note that $6 = \frac{6}{1}$. Multiply the numerator and denominator of $\frac{6}{1}$ by 3, but leave $\frac{8}{3}$ the way it is.)

20 $\frac{15}{36}$ and $\frac{10}{36}$ (Multiply the numerator and denominator of $\frac{5}{12}$ by 3, and multiply the numerator and denominator of $\frac{5}{18}$ by 2.

If your denominators are 72, 108, 144, 180, or 216, your denominator isn't the lowest common denominator: Make 36 instead, or reduce your answers to a denominator of 36.)

Chapter 4 – Comparing Fractions

Note: In the notes that follow, "common denom." refers to the method of finding a common denominator to compare fractions, whereas "cross mult." refers to the method of cross multiplying to compare fractions. The numbers for each case are provided in order to help you check your intermediate answers.

Part A

❶ $\frac{1}{2} > \frac{3}{7}$ (Common denom.: $\frac{7}{14} > \frac{6}{14}$. Cross mult.: $7 > 6$.)

❷ $\frac{6}{5} < \frac{3}{2}$ (Common denom.: $\frac{12}{10} < \frac{15}{10}$. Cross mult.: $12 < 15$.)

❸ $\frac{7}{9} > \frac{5}{8}$ (Common denom.: $\frac{56}{72} > \frac{45}{72}$. Cross mult.: $56 > 45$.)

❹ $\frac{7}{3} > 2$ (Common denom.: $\frac{7}{3} > \frac{6}{3}$. Cross mult.: $7 > 6$. Note that $2 = \frac{2}{1} = \frac{6}{3}$.)

❺ $\frac{3}{4} = \frac{12}{16}$ (Common denom.: $\frac{12}{16} = \frac{12}{16}$. Cross mult.: $48 = 48$.)

Part B

❻ $\frac{2}{7} < \frac{3}{8}$ (Common denom.: $\frac{16}{56} < \frac{21}{56}$. Cross mult.: $16 < 21$.)

❼ $\frac{4}{3} > \frac{5}{4}$ (Common denom.: $\frac{16}{12} > \frac{15}{12}$. Cross mult.: $16 > 15$.)

❽ $\frac{5}{6} > \frac{7}{9}$ (Common denom.: $\frac{15}{18} > \frac{14}{18}$. Cross mult.: $45 > 42$.)

❾ $\frac{5}{2} = \frac{15}{6}$ (Common denom.: $\frac{15}{6} = \frac{15}{6}$. Cross mult.: $30 = 30$.)

❿ $1 < \frac{10}{9}$ (Common denom.: $\frac{9}{9} < \frac{10}{9}$. Cross mult.: $9 < 10$. Note that $1 = \frac{1}{1} = \frac{9}{9}$.)

Part C

11 $\frac{3}{5} < \frac{7}{9}$ (Common denom.: $\frac{27}{45} < \frac{35}{45}$. Cross mult.: $27 < 35$.)

12 $\frac{1}{6} > \frac{1}{8}$ (Common denom.: $\frac{4}{24} > \frac{3}{24}$. Cross mult.: $8 > 6$.)

13 $\frac{3}{4} < \frac{8}{9}$ (Common denom.: $\frac{27}{36} < \frac{32}{36}$. Cross mult.: $27 < 32$.)

14 $\frac{6}{5} < \frac{4}{3}$ (Common denom.: $\frac{18}{15} < \frac{20}{15}$. Cross mult.: $18 < 20$.)

15 $\frac{12}{4} = 3$ (Common denom.: $\frac{12}{4} = \frac{12}{4}$. Cross mult.: $12 = 12$. Note that $3 = \frac{3}{1} = \frac{12}{4}$.)

Part D

16 $\frac{19}{6} > \frac{22}{7}$ (Common denom.: $\frac{133}{42} > \frac{132}{42}$. Cross mult.: $133 > 132$.)

17 $\frac{11}{3} < 4$ (Common denom.: $\frac{11}{3} < \frac{12}{3}$. Cross mult.: $11 < 12$. Note that $4 = \frac{4}{1} = \frac{12}{3}$.)

18 $\frac{10}{8} = \frac{15}{12}$ (Common denom.: $\frac{30}{24} = \frac{30}{24}$. Cross mult.: $120 = 120$.)

19 $\frac{18}{5} > \frac{18}{7}$ (Common denom.: $\frac{126}{35} > \frac{90}{35}$. Cross mult.: $126 > 90$.)

20 $\frac{1}{4} > \frac{3}{16}$ (Common denom.: $\frac{4}{16} > \frac{3}{16}$. Cross mult.: $16 > 12$.)

Chapter 5 – Adding Fractions

Part A

❶ $\frac{7}{12}$ (Multiply the numerator and denominator of $\frac{1}{3}$ by 4,

and multiply the numerator and denominator of $\frac{1}{4}$ by 3.)

❷ $\frac{22}{15}$ (Multiply the numerator and denominator of $\frac{2}{3}$ by 5,

and multiply the numerator and denominator of $\frac{4}{5}$ by 3.)

❸ $\frac{3}{8}$ (Multiply the numerator and denominator of $\frac{1}{4}$ by 2,

but leave $\frac{1}{8}$ the way it is.

If the denominator of your answer is 16, 24, or 32, you need to reduce your answer using the method from Chapter 2.)

❹ $\frac{41}{14}$ (Multiply the numerator and denominator of $\frac{3}{7}$ by 2,

and multiply the numerator and denominator of $\frac{5}{2}$ by 7.)

❺ $\frac{21}{4}$ (Note that $3 = \frac{3}{1}$. Multiply the numerator and denominator

of $\frac{3}{1}$ by 4, but leave $\frac{9}{4}$ the way it is. You should get $9 + 12$ in the numerator.)

Part B

❻ $\frac{23}{18}$ (Multiply the numerator and denominator of $\frac{5}{6}$ by 3,

and multiply the numerator and denominator of $\frac{4}{9}$ by 2.

If the denominator of your answer is 36 or 54, you need to

reduce your answer using the method from Chapter 2.)

❼ $\frac{112}{33}$ (Multiply the numerator and denominator of $\frac{8}{3}$ by 11,

and multiply the numerator and denominator of $\frac{8}{11}$ by 3.

You should get 88 + 24 in the numerator.)

❽ $\frac{85}{12}$ (Multiply the numerator and denominator of $\frac{15}{4}$ by 3,

and multiply the numerator and denominator of $\frac{10}{3}$ by 4.

You should get 45 + 40 in the numerator.)

❾ $\frac{12}{5}$ (Note that $1 = \frac{1}{1}$. Multiply the numerator and denominator

of $\frac{1}{1}$ by 5, but leave $\frac{7}{5}$ the way it is.)

❿ 1 (Note that $\frac{7}{11}$ and $\frac{4}{11}$ already have a common denominator.

Don't change either fraction: Simply add their numerators.

Note that $\frac{11}{11} = 11 \div 11 = 1$.)

Part C

⑪ $\frac{23}{36}$ (Multiply the numerator and denominator of $\frac{2}{9}$ by 4,

and multiply the numerator and denominator of $\frac{5}{12}$ by 3.

If the denominator of your answer is 72 or 108, you need to

reduce your answer using the method from Chapter 2.)

⑫ $\frac{25}{12}$ (Multiply the numerator and denominator of $\frac{4}{3}$ by 4,

and multiply the numerator and denominator of $\frac{3}{4}$ by 3.

Be careful in the numerator: You should get $16 + 9$, not $12 + 12$.)

⑬ $\frac{7}{6}$ (Multiply the numerator and denominator of $\frac{1}{2}$ by 3,

and multiply the numerator and denominator of $\frac{2}{3}$ by 2.

Be careful in the numerator: You should get $3 + 4$, not $3 + 2$.)

⑭ $\frac{71}{30}$ (Multiply the numerator and denominator of $\frac{7}{6}$ by 5,

and multiply the numerator and denominator of $\frac{6}{5}$ by 6.

Be careful in the numerator: You should get $35 + 36$, not $42 + 30$.)

⑮ $\frac{31}{4}$ (Note that $4 = \frac{4}{1}$. Multiply the numerator and denominator

of $\frac{4}{1}$ by 4, but leave $\frac{15}{4}$ the way it is. You should get $15 + 16$ in the

numerator.)

Part D

⑯ $\frac{127}{72}$ (Multiply the numerator and denominator of $\frac{7}{8}$ by 9,

and multiply the numerator and denominator of $\frac{8}{9}$ by 8.

Be careful in the numerator: You should get $63 + 64$, not $56 + 72$.)

⑰ $\frac{11}{16}$ (Multiply the numerator and denominator of $\frac{5}{8}$ by 2,

but leave $\frac{1}{16}$ the way it is. You should get $1 + 10$ in the numerator.

If the denominator of your answer is 32 or higher, you

need to reduce your answer using the method from Chapter 2.)

⑱ $\frac{58}{35}$ (Multiply the numerator and denominator of $\frac{4}{5}$ by 7,

and multiply the numerator and denominator of $\frac{6}{7}$ by 5.

Be careful in the numerator: You should get $28 + 30$, not $20 + 42$.)

⑲ $\frac{55}{7}$ (Note that $5 = \frac{5}{1}$. Multiply the numerator and denominator

of $\frac{5}{1}$ by 5, but leave $\frac{20}{7}$ the way it is. You should get $35 + 20$ in the

numerator.)

⑳ $\frac{1}{2}$ (Note that $\frac{1}{4}$ and $\frac{1}{4}$ already have a common denominator.

Don't change either fraction: Simply add their numerators.

If your answer is $\frac{2}{4}$, you need to reduce your answer using the

method from Chapter 2.)

Chapter 6 – Subtracting Fractions

Part A

❶ $\frac{1}{12}$ (Multiply the numerator and denominator of $\frac{1}{3}$ by 4,

and multiply the numerator and denominator of $\frac{1}{4}$ by 3.)

❷ $\frac{7}{20}$ (Multiply the numerator and denominator of $\frac{3}{4}$ by 5,

and multiply the numerator and denominator of $\frac{2}{5}$ by 4.

You should get $15 - 8$ in the numerator.)

❸ $\frac{11}{12}$ (Multiply the numerator and denominator of $\frac{7}{4}$ by 3,

and multiply the numerator and denominator of $\frac{5}{6}$ by 2.

You should get $21 - 10$ in the numerator.

If your answer is $\frac{22}{24}$, you need to reduce your answer using the

method from Chapter 2.)

❹ $\frac{1}{3}$ (Multiply the numerator and denominator of $\frac{1}{2}$ by 3,

but leave $\frac{1}{6}$ the way it is. You should get $3 - 1$ in the numerator.

Note that $\frac{2}{6}$ reduces to $\frac{1}{3}$ using the method from Chapter 2.

If your answer is $\frac{4}{12}$, you need to reduce your answer using the

method from Chapter 2.)

❺ $\frac{5}{4}$ (Note that $3 = \frac{3}{1}$. Multiply the numerator and denominator

of $\frac{3}{1}$ by 4, but leave $\frac{17}{4}$ the way it is. You should get $17 - 12$ in the

numerator.)

Part B

❻ $\frac{5}{72}$ (Multiply the numerator and denominator of $\frac{4}{9}$ by 8,

and multiply the numerator and denominator of $\frac{3}{8}$ by 9.

You should get $32 - 27$ in the numerator.)

❼ $\frac{5}{24}$ (Multiply the numerator and denominator of $\frac{5}{6}$ by 4,

and multiply the numerator and denominator of $\frac{5}{8}$ by 3.

If your answer is $\frac{10}{48}$, you need to reduce your answer using the

method from Chapter 2.)

❽ $\frac{7}{12}$ (Multiply the numerator and denominator of $\frac{9}{4}$ by 3,

and multiply the numerator and denominator of $\frac{5}{3}$ by 4.

You should get $27 - 20$ in the numerator.)

❾ $\frac{2}{5}$ (Note that $1 = \frac{1}{1}$. Multiply the numerator and denominator

of $\frac{1}{1}$ by 5, but leave $\frac{3}{5}$ the way it is.)

❿ $\frac{13}{15}$ (Multiply the numerator and denominator of $\frac{11}{5}$ by 9,

and multiply the numerator and denominator of $\frac{12}{9}$ by 5.

You should get $99 - 60$ in the numerator.

Note that $\frac{39}{45}$ reduces to $\frac{13}{15}$ using the method from Chapter 2,

since 39 and 45 are both divisible by 3.)

Part C

⓫ $\frac{1}{36}$ (Multiply the numerator and denominator of $\frac{5}{12}$ by 3,

and multiply the numerator and denominator of $\frac{7}{18}$ by 2.

If the denominator of your answer is 72, 108, 144, 180, or 216, you

need to reduce your answer using the method from Chapter 2.)

⓬ $\frac{11}{60}$ (Multiply the numerator and denominator of $\frac{9}{20}$ by 3,

and multiply the numerator and denominator of $\frac{4}{15}$ by 4.

You should get $27 - 16$ in the numerator.

If the denominator of your answer is 120, 180, 240, or 300, you need

to reduce your answer using the method from Chapter 2.)

⓭ $\frac{3}{4}$ (Note that $2 = \frac{2}{1}$. Multiply the numerator and denominator

of $\frac{2}{1}$ by 4, but leave $\frac{5}{4}$ the way it is. You should get $8 - 5$ in the

numerator.)

⓮ $\frac{2}{3}$ (Note that $\frac{8}{9}$ and $\frac{2}{9}$ already have a common denominator.

Don't change either fraction: Simply subtract their numerators.

If your answer is $\frac{6}{9}$, you need to reduce your answer using the

method from Chapter 2.)

⓯ $\frac{9}{20}$ (Multiply the numerator and denominator of $\frac{5}{4}$ by 5,

and multiply the numerator and denominator of $\frac{4}{5}$ by 4.

You should get $25 - 16$ in the numerator.)

Part D

⑯ $\frac{59}{10}$ (Multiply the numerator and denominator of $\frac{15}{2}$ by 5,

and multiply the numerator and denominator of $\frac{8}{5}$ by 2.

You should get $75 - 16$ in the numerator.)

⑰ $\frac{31}{72}$ (Multiply the numerator and denominator of $\frac{11}{24}$ by 3,

and multiply the numerator and denominator of $\frac{1}{36}$ by 2.

You should get $33 - 2$ in the numerator.

If the denominator of your answer is 144 or higher,

you need to reduce your answer using the method from Chapter 2.)

⑱ $\frac{49}{20}$ (Multiply the numerator and denominator of $\frac{16}{5}$ by 4,

and multiply the numerator and denominator of $\frac{3}{4}$ by 5.

You should get $64 - 15$ in the numerator.)

⑲ $\frac{7}{18}$ (Multiply the numerator and denominator of $\frac{2}{9}$ by 2,

but leave $\frac{11}{18}$ the way it is. You should get $11 - 4$ in the numerator.

If the denominator of your answer is 36, 54, 72, 90, 108, 126,

144, or 162, you need to reduce your answer using the method

from Chapter 2.)

⑳ $\frac{3}{4}$ (Note that $3 = \frac{3}{1}$. Multiply the numerator and denominator

of $\frac{3}{1}$ by 4, but leave $\frac{9}{4}$ the way it is. You should get $12 - 9$ in the

numerator.)

Chapter 7 – Multiplying Fractions

Part A

❶ $\frac{3}{10}$

❷ $\frac{10}{9}$ (Note that $\frac{20}{18} = \frac{10}{9}$.)

❸ $\frac{5}{16}$ (Note that $\frac{15}{48} = \frac{5}{16}$.)

❹ $\frac{1}{16}$

❺ $\frac{16}{3}$ (Note that $4 = \frac{4}{1}$. Multiply $\frac{4}{3} \times \frac{4}{1}$.)

Part B

❻ $\frac{12}{5}$ (Note that $\frac{36}{15} = \frac{12}{5}$.)

❼ $\frac{8}{35}$

❽ $\frac{3}{2}$ (Note that $\frac{30}{20} = \frac{3}{2}$.)

❾ $\frac{7}{5}$ (Note that $1 = \frac{1}{1}$. Multiply $\frac{1}{1} \times \frac{7}{5}$.)

❿ $\frac{1}{2}$ (Note that $\frac{72}{144} = \frac{1}{2}$. Divide the numerator and denominator by 72. Alternatively, cancel the 9's to get $\frac{8}{16}$, and note that $\frac{8}{16} = \frac{1}{2}$.)

Part C

⑪ $\frac{2}{15}$ (Note that $\frac{12}{90} = \frac{2}{15}$. Divide the numerator and denominator by 6.)

⑫ $\frac{4}{9}$

⑬ $\frac{9}{4}$ (Note that $\frac{135}{60} = \frac{9}{4}$. Divide the numerator and denominator by 15.)

⑭ 1 (Note that $\frac{28}{28} = 1$ since $28 \div 28 = 1$.

Alternatively, note that the 7's and 4's both cancel out: $\frac{7}{7} = 1$ and $\frac{4}{4} = 1$.

Also note that $\frac{1}{1} = 1$.)

⑮ $\frac{10}{3}$ (Note that $4 = \frac{4}{1}$. Multiply $\frac{5}{6} \times \frac{4}{1}$. Also note that $\frac{20}{6} = \frac{10}{3}$.)

Chapter 8 – Reciprocals

Part A

❶ $\frac{3}{2}$

❺ $\frac{8}{5}$ (Check the problem number.)

❷ $\frac{5}{6}$ (Check the problem number.)

❻ $\frac{1}{2}$ (Note that $\frac{2}{1} = 2$.)

❸ 3 (Note that $\frac{3}{1} = 3$.)

❼ $\frac{10}{7}$

❹ $\frac{1}{9}$ (Note that $\frac{9}{1} = 9$.)

❽ 25 (Note that $\frac{25}{1} = 25$.)

Part B

❾ $\frac{2}{9}$

⓭ 4 (Note that $\frac{4}{1} = 4$.)

❿ $\frac{1}{5}$ (Note that $\frac{5}{1} = 5$.)

⓮ $\frac{4}{3}$

⓫ 200 (Note that $\frac{200}{1} = 200$.)

⓯ $\frac{16}{9}$

⓬ $\frac{15}{4}$

⓰ 1 (Note that $\frac{1}{1} = 1$.)

Chapter 9 – Dividing Fractions

Part A

❶ $\frac{12}{35}$ (First rewrite the problem using a reciprocal: $\frac{6}{7} \times \frac{2}{5}$.

If your answer is $\frac{15}{7}$ or $\frac{30}{14}$, you may have multiplied by $\frac{5}{2}$ instead of $\frac{2}{5}$.)

❷ $\frac{3}{2}$ (First rewrite the problem using a reciprocal: $\frac{3}{4} \times \frac{2}{1}$.

Note that $\frac{6}{4} = \frac{3}{2}$.

If your answer is $\frac{3}{8}$, you may have multiplied by $\frac{1}{2}$ instead of $\frac{2}{1}$.)

❸ $\frac{14}{5}$ (First rewrite the problem using a reciprocal: $\frac{6}{5} \times \frac{7}{3}$.

Note that $\frac{42}{15} = \frac{14}{5}$. Divide the numerator and denominator by 3.

Note that $42 \div 14 = 3$.

If your answer is $\frac{18}{35}$, you may have multiplied by $\frac{3}{7}$ instead of $\frac{7}{3}$.)

❹ $\frac{18}{5}$ (First replace 6 with $\frac{6}{1}$.

Next rewrite the problem using a reciprocal: $\frac{6}{1} \times \frac{3}{5}$.

If your answer is 10 or $\frac{30}{3}$, you may have multiplied by $\frac{5}{3}$ instead of $\frac{3}{5}$.)

❺ 2 (First rewrite the problem using a reciprocal: $\frac{1}{2} \times \frac{4}{1}$.

Note that $\frac{4}{2} = 4 \div 2 = 2$. Also note that $\frac{2}{1} = 2$.

If your answer is $\frac{1}{8}$, you may have multiplied by $\frac{1}{4}$ instead of $\frac{4}{1}$.)

Part B

❻ $\frac{9}{32}$ (First rewrite the problem using a reciprocal: $\frac{3}{8} \times \frac{3}{4}$.

If your answer is $\frac{1}{2}$ or $\frac{12}{24}$, you may have multiplied by $\frac{4}{3}$ instead of $\frac{3}{4}$.)

❼ $\frac{3}{5}$ (First rewrite the problem using a reciprocal: $\frac{4}{3} \times \frac{9}{20}$.

Note that $\frac{36}{60} = \frac{3}{5}$. Divide the numerator and denominator by 12.

Note that $36 \div 12 = 3$ and $60 \div 12 = 5$.

Also note that $\frac{6}{10} = \frac{3}{5}$.

If your answer is $\frac{80}{27}$, you may have multiplied by $\frac{20}{9}$ instead of $\frac{9}{20}$.)

❽ $\frac{6}{7}$ (First rewrite the problem using a reciprocal: $\frac{16}{7} \times \frac{3}{8}$.

Note that $\frac{48}{56} = \frac{6}{7}$. Divide the numerator and denominator by 8.

Note that $48 \div 8 = 6$ and $56 \div 8 = 7$.

If your answer is $\frac{128}{21}$, you may have multiplied by $\frac{8}{3}$ instead of $\frac{3}{8}$.)

❾ $\frac{3}{8}$ (First replace 2 with $\frac{2}{1}$.

Next rewrite the problem using a reciprocal: $\frac{3}{4} \times \frac{1}{2}$.

If your answer is $\frac{3}{2}$ or $\frac{6}{4}$, you may have multiplied by $\frac{2}{1}$ instead of $\frac{1}{2}$.)

❿ 5 (First rewrite the problem using a reciprocal: $\frac{5}{6} \times \frac{6}{1}$.

Note that $\frac{30}{6} = 30 \div 6 = 5$.

Alternatively, cancel the 6's to get $\frac{5}{1} = 5$.

If your answer is $\frac{5}{36}$, you may have multiplied by $\frac{1}{6}$ instead of $\frac{6}{1}$.)

Part C

⓫ $\frac{1}{12}$ (First rewrite the problem using a reciprocal: $\frac{3}{16} \times \frac{4}{9}$.

Note that $\frac{12}{144} = \frac{1}{12}$. Divide the numerator and denominator by 12.

Note that $12 \div 12 = 1$ and $144 \div 12 = 12$.

If your answer is $\frac{27}{64}$, you may have multiplied by $\frac{9}{4}$ instead of $\frac{4}{9}$.)

⓬ $\frac{4}{25}$ (First rewrite the problem using a reciprocal: $\frac{2}{5} \times \frac{2}{5}$.

If your answer is 1 or $\frac{10}{10}$, you may have multiplied by $\frac{5}{2}$ instead of $\frac{2}{5}$.)

⓭ $\frac{5}{3}$ (First rewrite the problem using a reciprocal: $\frac{6}{5} \times \frac{25}{18}$.

Note that $\frac{150}{90} = \frac{5}{3}$. Divide the numerator and denominator by 30.

Note that $150 \div 30 = 5$ and $90 \div 30 = 3$.

If your answer is $\frac{108}{125}$, you may have multiplied by $\frac{18}{25}$ instead of $\frac{25}{18}$.)

⓮ 1 (First rewrite the problem using a reciprocal: $\frac{5}{8} \times \frac{8}{5}$.

Note that $\frac{40}{40} = 1$ since $40 \div 40 = 1$.

Alternatively, note that the 5's and 8's both cancel out: $\frac{5}{5} = 1$ and $\frac{8}{8} = 1$.

Also note that $\frac{1}{1} = 1$.

If your answer is $\frac{25}{64}$, you may have multiplied by $\frac{5}{8}$ instead of $\frac{8}{5}$.)

⓯ 12 (First replace 8 with $\frac{8}{1}$.

Next rewrite the problem using a reciprocal: $\frac{8}{1} \times \frac{3}{2}$.

Note that $\frac{24}{2} = 24 \div 2 = 12$. Also note that $\frac{12}{1} = 12$.

If your answer is $\frac{16}{3}$, you may have multiplied by $\frac{2}{3}$ instead of $\frac{3}{2}$.)

Chapter 10 – Decimals

Part A

❶ $\frac{1}{2}$ (If your answer is $\frac{5}{10}$, you need to reduce your answer using the method from Chapter 2.)

❷ $\frac{3}{4}$ (If your answer is $\frac{75}{100}$, you need to reduce your answer using the method from Chapter 2.)

❸ $\frac{11}{10}$

❹ $\frac{13}{40}$ (Note that the 5 of 0.325 is in the thousandths place. If your answer is $\frac{325}{1000}$, you need to reduce your answer using the method from Chapter 2. Note that $325 \div 25 = 13$ and $1000 \div 25 = 40$.)

❺ $\frac{7}{4}$ (If your answer is $\frac{175}{100}$, you need to reduce your answer using the method from Chapter 2.)

Part B

❻ $\frac{9}{10}$

❼ $\frac{1}{20}$ (If your answer is $\frac{5}{100}$, you need to reduce your answer using the method from Chapter 2.)

❽ $\frac{5}{2}$ (If your answer is $\frac{25}{10}$, you need to reduce your answer using the method from Chapter 2. Divide the numerator and denominator by 5.)

❾ $\frac{1}{500}$ (Note that the 2 of 0.002 is in the thousandths place. If your answer is $\frac{2}{1000}$, you need to reduce your answer using the method from Chapter 2.)

❿ $\frac{6}{5}$ (If your answer is $\frac{12}{100}$, you need to reduce your answer using the method from Chapter 2.)

Part C

⓫ 1.5 (Multiply the numerator and denominator of $\frac{3}{2}$ by 5.

You should get $\frac{15}{10}$, which equals 1.5.)

⓬ 0.6 (Multiply the numerator and denominator of $\frac{3}{5}$ by 2.

You should get $\frac{6}{10}$, which equals 0.6.

Note that .6 is the same as 0.6. However, .06 is incorrect.)

⓭ 0.55 (Multiply the numerator and denominator of $\frac{11}{20}$ by 5.

You should get $\frac{55}{100}$, which equals 0.55.)

⓮ 1.25 (Multiply the numerator and denominator of $\frac{5}{4}$ by 25.

You should get $\frac{125}{100}$, which equals 1.25.)

⓯ 0.028 (Multiply the numerator and denominator of $\frac{7}{250}$ by 4.

You should get $\frac{28}{1000}$, which equals 0.028. Note that the 8 must be

in the thousandths place since $\frac{28}{1000}$ is over 1000.

Therefore, an answer of 0.28 would be incorrect.)

Part D

16 1.42 (Multiply the numerator and denominator of $\frac{71}{50}$ by 2. You should get $\frac{142}{100}$, which equals 1.42.)

17 1.7 (Don't multiply by anything since $\frac{17}{10}$ is already over a power of 10. Note that $\frac{17}{10}$ equals 1.7.)

18 0.64 (Multiply the numerator and denominator of $\frac{16}{25}$ by 4. You should get $\frac{64}{100}$, which equals 0.64.)

19 0.375 (Multiply the numerator and denominator of $\frac{3}{8}$ by 125. Note that $8 \times 125 = 1000$. You should get $\frac{375}{1000}$, which equals 0.375. Note that the 5 must be in the thousandths place since $\frac{375}{1000}$ is over 1000.)

20 0.045 (Multiply the numerator and denominator of $\frac{9}{200}$ by 5. You should get $\frac{45}{1000}$, which equals 0.045. Note that the 5 must be in the thousandths place since $\frac{45}{1000}$ is over 1000. Therefore, an answer of 0.45 would be incorrect.)

Chapter 11 – Repeating Decimals

Part A

❶ $\frac{2}{3}$ (Reduce $\frac{6}{9}$ using the method from Chapter 2.

If your answer is $\frac{3}{5}$ or $\frac{6}{10}$, you read the problem as 0.6 instead of $0.\bar{6}$.

The distinction is that $0.\bar{6} = 0.66666666...$

You can enter $2 \div 3$ on a calculator to check that $\frac{2}{3} = 0.66666666...$

Note that a calculator will round the last 6 up to a 7.)

❷ $\frac{7}{11}$ (Since two digits are repeating, divide 63 by 99.

Note that $\frac{63}{99}$ reduces to $\frac{7}{11}$. Also note that $63 \div 9 = 7$ and $99 \div 9 = 11$.

If your answer is $\frac{63}{100}$, you read the problem as 0.63 instead of $0.\overline{63}$.

You can enter $7 \div 11$ on a calculator to check that $\frac{7}{11} = 0.63636363...$)

❸ $\frac{32}{99}$ (Since two digits are repeating, divide 32 by 99.

If your answer is $\frac{8}{25}, \frac{16}{50}$, or $\frac{32}{100}$, you read the problem as 0.32 instead of $0.\overline{32}$.

You can enter $32 \div 99$ on a calculator to check that $\frac{32}{99} = 0.32323232...$)

❹ $\frac{4}{45}$ (Your intermediate answer should be $\frac{8}{90}$. Divide $\frac{8}{9}$ by 10 to get $\frac{8}{90}$

because there is one 0 between the decimal point and the 8 in $0.0\bar{8}$.

Reduce $\frac{8}{90}$ using the method from Chapter 2. Note that $\frac{8}{90} = \frac{4}{45}$.

If your answer is $\frac{2}{25}, \frac{4}{50}$, or $\frac{8}{100}$, you read the problem as 0.08 instead of $0.0\bar{8}$.

You can enter $4 \div 45$ on a calculator to check that $\frac{4}{45} = 0.088888888...$

Note that a calculator will round the last 8 up to a 9.)

❺ $\frac{592}{999}$ (Since three digits are repeating, divide 592 by 999.

If your answer is $\frac{74}{125}$ or $\frac{592}{1000}$, you read the problem as 0.592 instead of $0.\overline{592}$.

You can enter $592 \div 999$ on a calculator to check that $\frac{592}{999} = 0.592592592...$)

Part B

❻ $\frac{4}{3}$ (Separate $1.\overline{3}$ into nonrepeating and repeating parts: $1.\overline{3} = 1 + 0.\overline{3}$.

Note that $1 = \frac{1}{1}$ and $0.\overline{3} = \frac{3}{9}$. Note that $\frac{3}{9}$ reduces to $\frac{1}{3}$.

Multiply both the numerator and denominator of $\frac{1}{1}$ by 3 in order to

make a common denominator of 3. You should get $\frac{3}{3}$.

Add the fractions: $1 + \frac{1}{3} = \frac{1}{1} + \frac{1}{3} = \frac{3}{3} + \frac{1}{3} = \frac{4}{3}$.

If your answer is $\frac{13}{10}$, you read the problem as 1.3 instead of $1.\overline{3}$.

You can enter $4 \div 3$ on a calculator to check that $\frac{4}{3} = 1.33333333...$)

❼ $\frac{1}{60}$ (Separate $0.01\overline{6}$ into nonrepeating and repeating parts:

$0.01\overline{6} = 0.01 + 0.00\overline{6}$. The 6 is in the thousandths place.

Note that $0.01 = \frac{1}{100}$ since the 1 is in the hundredths place.

Note that $0.00\overline{6} = \frac{6}{900}$. Divide $\frac{6}{9}$ by 100 to get $\frac{6}{900}$ because there

are two 0's between the decimal point and the 6 in $0.00\overline{6}$.

Multiply both the numerator and denominator of $\frac{1}{100}$ by 9 in order to

make a common denominator of 900. You should get $\frac{9}{900}$.

Add the fractions: $\frac{1}{100} + \frac{6}{900} = \frac{9}{900} + \frac{6}{900} = \frac{15}{900}$.

Note that $\frac{15}{900}$ reduces to $\frac{1}{60}$. Also note that $15 \div 15 = 1$ and $900 \div 15 = 60$.

If your answer is $\frac{3}{180}, \frac{5}{300}$, or $\frac{15}{900}$, you need to reduce your answer.

If your answer is $\frac{8}{495}$ or $\frac{16}{990}$, you read the problem as $0.0\overline{16}$ instead of $0.01\overline{6}$.

If your answer is $\frac{2}{125}$ or $\frac{16}{1000}$, you read the problem as 0.016 instead of $0.01\overline{6}$.

You can enter $1 \div 60$ on a calculator to check that $\frac{1}{60} = 0.0166666666...$

Note that a calculator will round the last 6 up to a 7.)

❽ $\frac{7}{30}$ (Separate $0.2\bar{3}$ into nonrepeating and repeating parts:

$0.2\bar{3} = 0.2 + 0.0\bar{3}$. Note that $0.2 = \frac{2}{10}$ and $0.0\bar{3} = \frac{3}{90}$. Divide $\frac{3}{9}$ by 10 to

get $\frac{3}{90}$ because there is one 0 between the decimal point and the 3 in $0.0\bar{3}$.

Note that $\frac{3}{90}$ reduces to $\frac{1}{30}$. Multiply both the numerator and denominator

of $\frac{2}{10}$ by 3 in order to make a common denominator of 30. You should get $\frac{6}{30}$.

Add the fractions: $\frac{2}{10} + \frac{3}{90} = \frac{2}{10} + \frac{1}{30} = \frac{6}{30} + \frac{1}{30} = \frac{7}{30}$.

If your answer is $\frac{23}{99}$, you read the problem as $0.\overline{23}$ instead of $0.2\bar{3}$.

If your answer is $\frac{23}{100}$, you read the problem as 0.23 instead of $0.2\bar{3}$.

You can enter $7 \div 30$ on a calculator to check that $\frac{7}{30} = 0.233333333...$)

❾ $\frac{25}{999}$ (Since three digits are repeating, divide 25 by 999.

If your answer is $\frac{5}{198}$ or $\frac{25}{990}$, you read the problem as $0.0\overline{25}$ instead of $0.\overline{025}$.

If your answer is $\frac{1}{40}$ or $\frac{25}{1000}$, you read the problem as 0.025 instead of $0.\overline{025}$.

You can enter $25 \div 999$ on a calculator to check that $\frac{25}{999} = 0.025025025...$)

❿ $\frac{481}{990}$ (Separate $0.4\overline{85}$ into nonrepeating and repeating parts:

$0.4\overline{85} = 0.4 + 0.0\overline{85}$. The 5 is in the thousandths place.

Note that $0.4 = \frac{4}{10}$ since the 4 is in the tenths place.

Note that $0.0\overline{85} = \frac{85}{990}$. Divide $\frac{85}{99}$ by 10 to get $\frac{85}{990}$ because there

is one 0 between the decimal point and the 8 in $0.0\overline{85}$.

Multiply both the numerator and denominator of $\frac{4}{10}$ by 99 in order to

make a common denominator of 990. You should get $\frac{396}{990}$.

Add the fractions: $\frac{4}{10} + \frac{85}{990} = \frac{396}{990} + \frac{85}{990} = \frac{481}{990}$.

If your answer is $\frac{485}{999}$, you read the problem as $0.\overline{485}$ instead of $0.4\overline{85}$.

If your answer is $\frac{97}{200}$ or $\frac{485}{1000}$, you read the problem as 0.485 instead of $0.4\overline{85}$.

You can enter $481 \div 990$ on a calculator to check that $\frac{481}{990} = 0.485858585...$

Note that a calculator will round the last digit up.)

Part C

⓫ $0.\overline{6}$ (Multiply the numerator and denominator of $\frac{2}{3}$ by 3.

You should get $\frac{6}{9}$. None of the special cases apply.

The only digit of the numerator repeats: $\frac{6}{9} = 0.\overline{6}$.

Look closely: 0.6 is incorrect. Put a bar over the 6 in $0.\overline{6}$.

You can enter $2 \div 3$ on a calculator to check that $\frac{2}{3} = 0.66666666...$

Note that a calculator will round the last 6 up to a 7.)

⓬ $0.00\overline{3}$ (Multiply the numerator and denominator of $\frac{1}{300}$ by 3.

You should get $\frac{3}{900}$. Since the denominator has 0's, this is one of the

special cases. First write $\frac{3}{9} = 0.\overline{3}$. Next shift the decimal point two

places to the left to make up for the 0's that were removed from

$\frac{3}{900}$ to make $\frac{3}{9}$. You should get $\frac{3}{900} = 0.00\overline{3}$.

Look closely: 0.003 is incorrect. Put a bar over the 3 in $0.00\overline{3}$.

You can enter $1 \div 300$ on a calculator to check that $\frac{1}{3} = 0.0033333333...$)

⓭ $0.\overline{900}$ (Multiply the numerator and denominator of $\frac{100}{111}$ by 9.

You should get $\frac{900}{999}$. None of the special cases apply.

All three digits of the numerator repeat: $\frac{900}{999} = 0.\overline{900}$.

Look closely: 0.9, $0.\overline{9}$., and $0.\overline{90}$. are all incorrect.

Put a bar over the 900 in $0.\overline{900}$. All three digits repeat.

If you have $0.9\overline{009}$ or $0.90\overline{090}$, they are okay, but not as simple as $0.\overline{900}$.

However, $0.\overline{90}$, $0.\overline{9009}$, and $0.\overline{9090}$ are all incorrect.

You can enter $100 \div 111$ on a calculator to check that $\frac{100}{111} = 0.900900900...$)

⑭ $2.\overline{6}$ (Multiply the numerator and denominator of $\frac{8}{3}$ by 3.

You should get $\frac{24}{9}$. Since the numerator is larger than the denominator,

this is one of the special cases. Rewrite $\frac{24}{9}$ as a whole number plus a

fraction that is smaller than one. Note that $24 \div 9$ equals 2 with a

remainder of 6 (since $9 \times 2 = 18$ and $24 - 18 = 6$). Thus, $\frac{24}{9} = 2 + \frac{6}{9}$.

That's the whole number plus the remainder divided by 9.

If you would like more help with the previous steps, read Chapter 16.

Note that $\frac{6}{9} = 0.\overline{6}$. Plug this into $\frac{24}{9} = 2 + \frac{6}{9}$ to get $\frac{24}{9} = 2 + 0.\overline{6} = 2.\overline{6}$.

Look closely: 2.6 is incorrect. Put a bar over the 6 in $2.\overline{6}$.

You can enter $8 \div 3$ on a calculator to check that $\frac{8}{3} = 2.66666666...$

Note that a calculator will round the last 6 up to a 7.)

⑮ $0.\overline{006}$ (Multiply the numerator and denominator of $\frac{2}{333}$ by 3.

You should get $\frac{6}{999}$. Since the numerator has fewer digits than the

denominator, this is one of the special cases. There will be two

repeating 0's because the numerator has 2 fewer digits than the

denominator. Therefore, $\frac{6}{999} = 0.\overline{006}$.

Look closely: 0.006, $0.00\overline{6}$., and $0.0\overline{06}$. are all incorrect.

Put a bar over the 006 in $0.\overline{006}$. All three digits repeat.

If you have $0.0\overline{060}$ or $0.00\overline{600}$, they are okay, but not as simple as $0.\overline{006}$.

However, $0.0\overline{600}$, $0.\overline{06}$, and $0.00\overline{060}$ are all incorrect.

You can enter $2 \div 333$ on a calculator to check that $\frac{2}{333} = 0.006006006...$)

Part D

⓰ $1.\overline{1}$ (The denominator is already 9: Don't change it.

Since the numerator is larger than the denominator, this is one of the special cases. Rewrite $\frac{10}{9}$ as a whole number plus a fraction that is smaller than one. Note that $10 \div 9$ equals 1 with a remainder of 1 (since $9 \times 1 = 9$ and $10 - 1 = 9$). Thus, $\frac{10}{9} = 1 + \frac{1}{9}$.

That's the whole number plus the remainder divided by 9.

If you would like more help with the previous steps, read Chapter 16.

Note that $\frac{1}{9} = 0.\overline{1}$. Plug this into $\frac{10}{9} = 1 + \frac{1}{9}$ to get $\frac{10}{9} = 1 + 0.\overline{1} = 1.\overline{1}$.

Look closely: 1.1 is incorrect. Put a bar over the last 1 in $1.\overline{1}$.

You can enter $10 \div 9$ on a calculator to check that $\frac{10}{9} = 1.11111111...$)

⓱ $0.\overline{15}$ (Multiply the numerator and denominator of $\frac{5}{33}$ by 3.

You should get $\frac{15}{99}$. None of the special cases apply.

Both digits of the numerator repeat: $\frac{15}{99} = 0.\overline{15}$.

Look closely: 0.15 is incorrect. Put a bar over the 15 in $0.\overline{15}$.

Note that $0.1\overline{5}$ is also incorrect: You need the bar over both the 1 and 5.

You can enter $5 \div 33$ on a calculator to check that $\frac{5}{33} = 0.15151515...$)

⓲ $0.0\overline{8}$ (Multiply the numerator and denominator of $\frac{4}{45}$ by 2.

You should get $\frac{8}{90}$. Since the denominator has a 0, this is one of the special cases. First write $\frac{8}{9} = 0.\overline{8}$. Next shift the decimal point one place to the left for the 0 that was removed from $\frac{8}{90}$ to make $\frac{8}{9}$.

You should get $\frac{8}{90} = 0.0\overline{8}$.

Look closely: 0.08 is incorrect. Put a bar over the 8 in $0.0\overline{8}$.

You can enter $4 \div 45$ on a calculator to check that $\frac{4}{45} = 0.088888888......$

Note that a calculator will round the last 8 up to a 9.)

⑲ $0.5\bar{3}$ (Multiply the numerator and denominator of $\frac{8}{15}$ by 6.

You should get $\frac{48}{90}$. Since the denominator has a 0, this is one of the special cases. First ignore the 0 to get $\frac{48}{9}$. Now a second special case applies because the numerator of $\frac{48}{9}$ is larger than the denominator.

Rewrite $\frac{48}{9}$ as a whole number plus a fraction that is smaller than one. Note that $48 \div 9$ equals 5 with a remainder of 3 (since $9 \times 5 = 45$ and $48 - 45 = 3$). Thus, $\frac{48}{9} = 5 + \frac{3}{9}$. That's the whole number plus the remainder divided by 9. If you would like more help with the previous steps, read Chapter 16. Note that $\frac{3}{9} = 0.\bar{3}$.

Plug this into $\frac{48}{9} = 5 + \frac{3}{9}$ to get $\frac{48}{9} = 5 + 0.\bar{3} = 5.\bar{3}$.

Next shift the decimal point one place to the left for the 0 that was removed from $\frac{48}{90}$ to make $\frac{48}{9}$. You should get $\frac{48}{90} = 0.5\bar{3}$.

Look closely: 0.53 is incorrect. Put a bar over the 3 in $0.5\bar{3}$.

Note that $0.\overline{53}$ is also incorrect: The bar should be over the 3 only.

You can enter $8 \div 15$ on a calculator to check that $\frac{8}{15} = 0.533333333...$)

⑳ $0.\overline{077}$ (The denominator is already 999: Don't change it. Since the numerator has fewer digits than the denominator, this is one of the special cases. There will be a repeating 0 because the numerator has 1 less digit than the denominator has.

Therefore, $\frac{77}{999} = 0.\overline{077}$.

Look closely: 0.77, $0.0\bar{7}.$, and $0.0\overline{77}.$ are all incorrect.

Put a bar over the 077 in $0.\overline{077}$. All three digits repeat.

If you have $0.0\overline{770}$ or $0.07\overline{707}$, they are okay, but not as simple as $0.\overline{077}$.

However, $0.\overline{77}$, $0.0\overline{707}$, and $0.00\overline{770}$ are all incorrect.

You can enter $77 \div 999$ on a calculator to check that $\frac{77}{999} = 0.077077077...$)

Chapter 12 – Percentages

Part A

Note: You can confirm the following answers on a calculator by multiplying by 100.

❶ 70% (Shift the decimal point two places to the right.)

❷ 45% (Shift the decimal point two places to the right.)

❸ 110% (Shift the decimal point two places to the right.)

❹ 0.5% (Shift the decimal point two places to the right.
Note that 0.5% = 0.005 is smaller than 5% = 0.05.)

❺ 200% (Shift the decimal point two places to the right.)

Part B

Note: You can confirm the following answers on a calculator by dividing by 100.

❻ 0.2 (Shift the decimal point two places to the left.)

❼ 5 (Shift the decimal point two places to the left.)

❽ 0.025 (Shift the decimal point two places to the left.
Note that 2.5% = 0.025 is smaller than 25% = 0.25.)

❾ 3.25 (Shift the decimal point two places to the left.)

❿ 0.002 (Shift the decimal point two places to the left.
Note that 0.2% = 0.002 is smaller than 2% = 0.02.)

Part C

⓫ 75% (Multiply the numerator and denominator of $\frac{3}{4}$ by 25.

You should get $\frac{75}{100}$, which equals 0.75.

Shift the decimal point two places to the right to get 75%.

Enter $3 \div 4 \times 100$ on a calculator to verify that $\frac{3}{4} = 75\%$.)

⓬ 60% (Multiply the numerator and denominator of $\frac{3}{5}$ by 2.

You should get $\frac{6}{10}$, which equals 0.6.

Shift the decimal point two places to the right to get 60%.

Enter $3 \div 5 \times 100$ on a calculator to verify that $\frac{3}{5} = 60\%$.)

⓭ 32% (Multiply the numerator and denominator of $\frac{8}{25}$ by 4.

You should get $\frac{32}{100}$, which equals 0.32.

Shift the decimal point two places to the right to get 32%.

Enter $8 \div 25 \times 100$ on a calculator to verify that $\frac{8}{25} = 32\%$.)

⓮ 1.5% (Multiply the numerator and denominator of $\frac{3}{200}$ by 5.

You should get $\frac{15}{1000}$, which equals 0.015. Note that the 5 is in the

thousandths place because $\frac{15}{1000}$ is over 1000.

Shift the decimal point two places to the right to get 1.5%.

Enter $3 \div 200 \times 100$ on a calculator to verify that $\frac{3}{200} = 1.5\%$.)

⓯ 125% (Multiply the numerator and denominator of $\frac{5}{4}$ by 25.

You should get $\frac{125}{100}$, which equals 1.25.

Note that the 5 is in the hundredths place because $\frac{125}{100}$ is over 100.

Shift the decimal point two places to the right to get 125%.

Enter $5 \div 4 \times 100$ on a calculator to verify that $\frac{5}{4} = 125\%$.)

Part D

Note: An alternate method is to write the given number over 100 (see the alternate solution on the bottom of page 110).

⑯ $\frac{1}{2}$ (First shift the decimal point two places to the left to get 0.5.

Then write 0.5 as $\frac{5}{10}$. Finally, reduce your answer using the method from Chapter 2.

Enter $1 \div 2 \times 100$ on a calculator to verify that $\frac{1}{2} = 50\%$.)

⑰ $\frac{17}{20}$ (First shift the decimal point two places to the left to get 0.85.

Then write 0.85 as $\frac{85}{100}$. Finally, reduce your answer using the method from Chapter 2. Note that $85 \div 5 = 17$.

Enter $17 \div 20 \times 100$ on a calculator to verify that $\frac{17}{20} = 85\%$.)

⑱ $\frac{3}{2}$ (First shift the decimal point two places to the left to get 1.5.

Then write 1.5 as $\frac{15}{10}$. We divided by 10 because the 5 is in the tenths

position in the decimal 1.5. Finally, note that $\frac{15}{10} = \frac{15 \div 5}{10 \div 5} = \frac{3}{2}$.

Enter $3 \div 2 \times 100$ on a calculator to verify that $\frac{3}{2} = 150\%$.)

⑲ $\frac{1}{4}$ (First shift the decimal point two places to the left to get 0.25.

Then write 0.25 as $\frac{25}{100}$. Finally, reduce your answer using the method from Chapter 2. Divide the numerator and denominator by 25.

Enter $1 \div 4 \times 100$ on a calculator to verify that $\frac{1}{4} = 25\%$.)

⑳ $\frac{2}{25}$ (First shift the decimal point two places to the left to get 0.08.

Then write 0.08 as $\frac{8}{100}$. Finally, reduce your answer using the method from Chapter 2. Divide the numerator and denominator by 4.

Enter $2 \div 25 \times 100$ on a calculator to verify that $\frac{2}{25} = 8\%$.)

Chapter 13 – Adding and Subtracting Decimals

Note: You should be fluent in ordinary multi-digit addition and subtraction before attempting to add or subtract decimals. For example, you should know how to carry over when adding $14 + 19$ to make 33 or how to borrow when subtracting $31 - 12$ to make 9. If not, you should review multi-digit addition and subtraction first.

❶
$$\begin{array}{r} 5.2 \\ + 0.6 \\ \hline 5.8 \end{array}$$

❷
$$\begin{array}{r} 3.02 \\ + 0.90 \\ \hline 3.92 \end{array}$$

❸
$$\begin{array}{r} 0.060 \\ + 0.038 \\ \hline 0.098 \end{array}$$

❹
$$\begin{array}{r} 0.024 \\ + 0.018 \\ \hline 0.042 \end{array}$$
$(24 + 18 = 42)$

❺
$$\begin{array}{r} 15.6 \\ + 8.0 \\ \hline 23.6 \end{array}$$
$(15 + 8 = 23)$

❻
$$\begin{array}{r} 4.7 \\ + 5.9 \\ \hline 10.6 \end{array}$$
$(7 + 9 = 16$
$4 + 5 + 1 = 10)$

❼
$$\begin{array}{r} 0.800 \\ + 0.375 \\ \hline 1.175 \end{array}$$
$(8 + 3 = 11)$

❽
$$\begin{array}{r} 0.909 \\ + 0.001 \\ \hline 0.910 \end{array}$$
$(9 + 1 = 10$
same as $0.91)$

❾
$$\begin{array}{r} 0.36 \\ - 0.24 \\ \hline 0.12 \end{array}$$
(same as .12)

❿
$$\begin{array}{r} 0.854 \\ - 0.030 \\ \hline 0.824 \end{array}$$

⓫
$$\begin{array}{r} 0.089 \\ - 0.007 \\ \hline 0.082 \end{array}$$

⓬
$$\begin{array}{r} 4.20 \\ - 1.75 \\ \hline 2.45 \end{array}$$
(Borrow:
$10 - 5 = 5$
$11 - 7 = 4$
$3 - 1 = 2)$

⓭
$$\begin{array}{r} 0.500 \\ - 0.125 \\ \hline 0.375 \end{array}$$
(Borrow:
$10 - 5 = 5$
$9 - 2 = 7$
$4 - 1 = 3)$

⓮
$$\begin{array}{r} 0.072 \\ - 0.038 \\ \hline 0.034 \end{array}$$
$(72 - 38 = 34)$

⓯
$$\begin{array}{r} 0.762 \\ - 0.400 \\ \hline 0.362 \end{array}$$

⓰
$$\begin{array}{r} 3.000 \\ - 0.025 \\ \hline 2.975 \end{array}$$
(Borrow:
$10 - 5 = 5$
$9 - 2 = 7$
$9 - 0 = 9$
$2 - 0 = 2)$

Note: You can check your answer with a calculator. For example, $3 - 0.025 = 2.975$.

Chapter 14 – Multiplying and Dividing Decimals

Note: You should be fluent in ordinary multi-digit multiplication and division before attempting to multiply or divide decimals. If not, you should review multi-digit multiplication and division first.

Part A

❶
$$\begin{array}{r} 0.9 \\ \times\, 0.3 \\ \hline 0.27 \end{array}$$
(or .27)

❷
$$\begin{array}{r} 1.2 \\ \times\, 0.4 \\ \hline 0.48 \end{array}$$
(or .48)

❸
$$\begin{array}{r} 0.05 \\ \times\, 0.6 \\ \hline 0.030 \end{array}$$
(same as 0.03)

❹
$$\begin{array}{r} 0.003 \\ \times\, 0.02 \\ \hline 0.00006 \end{array}$$

❺
$$\begin{array}{r} 1.2 \\ \times\, 1.2 \\ \hline 1.44 \end{array}$$

❻
$$\begin{array}{r} 1.75 \\ \times\, 4 \\ \hline 7.00 \end{array}$$
(same as 7)

❼
$$\begin{array}{r} 2.25 \\ \times\, 0.5 \\ \hline 1.125 \end{array}$$

❽
$$\begin{array}{r} 0.09 \\ \times\, 0.04 \\ \hline 0.0036 \end{array}$$

❾
$$\begin{array}{r} 100 \\ \times\, 0.7 \\ \hline 70.0 \end{array}$$
(same as 70)

❿
$$\begin{array}{r} 0.025 \\ \times\, 0.08 \\ \hline 0.00200 \end{array}$$
(same as 0.002)

⓫
$$\begin{array}{r} 3000 \\ \times\, 0.04 \\ \hline 120.00 \end{array}$$
(same as 120)

⓬
$$\begin{array}{r} 0.0001 \\ \times\, 0.001 \\ \hline 0.0000001 \end{array}$$

Part B

⓭ 4 (Multiply 0.48 and 0.12 both by 100, and express them as a fraction. You should get $\frac{48}{12}$, which reduces to $\frac{4}{1}$.

Note that $\frac{4}{1} = 4 \div 1 = 4$ or note that $\frac{48}{12} = 48 \div 12 = 4$.

An answer of 4.0 or 4.00 is also correct, though it's customary to ignore trailing zeros.)

⓮ 0.75 (Multiply 0.6 and 0.8 both by 10, and express them as a fraction. You should get $\frac{6}{8}$, which reduces to $\frac{3}{4}$.

Multiply the numerator and denominator of $\frac{3}{4}$ by 25.

You should get $\frac{75}{100}$, which equals 0.75. It's the same as .75)

⓯ 450 (Multiply 4.5 and 0.01 both by 100.

Note that $4.5 \times 100 = 450$ and $0.01 \times 100 = 1$.

Express them as a fraction. You should get $\frac{450}{1}$.

Note that $\frac{450}{1} = 450 \div 1 = 450$.

Anything divided by 1 equals itself.)

⓰ 1.5 (Multiply 0.75 and 0.5 both by 100, and express them as a fraction. You should get $\frac{75}{50}$, which reduces to $\frac{3}{2}$ if you divide the numerator and denominator both by 25.

Multiply the numerator and denominator of $\frac{3}{2}$ by 5.

You should get $\frac{15}{10}$, which equals 1.5.

An answer of 1.50 is also correct, though it's customary to ignore trailing zeros.)

Part C

❶⓱ 0.2 (Multiply 0.08 and 0.4 both by 100, and express them as a fraction. You should get $\frac{8}{40}$, which reduces to $\frac{1}{5}$ if you divide the numerator and denominator both by 8. Note that $40 \div 8 = 5$. Multiply the numerator and denominator of $\frac{1}{5}$ by 2. You should get $\frac{2}{10}$, which equals 0.2. It's the same as .2. An answer of 0.20 is also correct, though it's customary to ignore trailing zeros.)

⓲ 4 (Multiply 1 and 0.25 both by 100, and express them as a fraction. You should get $\frac{100}{25}$, which reduces to $\frac{4}{1}$. Note that $\frac{4}{1} = 4 \div 1 = 4$ or note that $\frac{100}{25} = 100 \div 25 = 4$. An answer of 4.0 or 4.00 is also correct, though it's customary to ignore trailing zeros.)

⓳ 0.15 (Multiply 0.048 and 0.32 both by 1000. Shift the decimal point three places to the right. Note that $0.048 \times 1000 = 48$ and $0.32 \times 1000 = 320$. Shift the decimal point two places to the right to go from 0.32 to 32, and then shift another place to the right to go from 32 to 320. Express them as a fraction. You should get $\frac{48}{320}$. Note that $\frac{48}{320}$ reduces to $\frac{3}{20}$ if you divide the numerator and denominator both by 16. Note that $48 \div 16 = 3$ and $320 \div 16 = 20$. Multiply the numerator and denominator of $\frac{3}{20}$ by 5. You should get $\frac{15}{100}$, which equals 0.15. It's the same as .15.)

⓴ 0.1 (Multiply 0.00032 and 0.0032 both by 10,000, and express them as a fraction. You should get $\frac{32}{320}$, which reduces to $\frac{1}{10}$ if you divide the numerator and denominator both be 32. Note that $320 \div 32 = 10$. Note that $\frac{1}{10}$ equals 0.1. It's the same as .1.)

Chapter 15 – Word Problems

Note: If you're struggling with one of these word problems, try to find an example that is very similar to the problem to help serve as a guide.

❶ $\frac{4}{5}$ (You should get $\frac{16}{20}$ as an intermediate answer.
Reduce your answer using the method from Chapter 2.)

❷ $\frac{2}{3}$ (Add the given numbers together to find the denominator.
You should get $\frac{6}{9}$ as an intermediate answer.
Reduce your answer using the method from Chapter 2.)

❸ $\frac{3}{7}$ (Add the given numbers together to find the denominator.
You should get $\frac{18}{42}$ as an intermediate answer.
Reduce your answer using the method from Chapter 2.
Note that $18 \div 6 = 3$ and $42 \div 6 = 7$.
If your answer is $\frac{4}{7}$, $\frac{12}{21}$, or $\frac{24}{42}$, you used yes votes instead of no votes for your numerator.)

❹ $\frac{4}{5}$ (Add the given numbers together to find the denominator.
You should get $\frac{16}{20}$ as an intermediate answer.
Reduce your answer using the method from Chapter 2.)

❺ 2:1 (You should get $\frac{12}{6}$ as an intermediate answer. Note that $12 \div 6 = 2$.
The ratio is 2 to 1 which we can write as 2:1. Note that 12:6 reduces to 2:1.)

❻ $\frac{29}{24}$ hours (Add the given fractions together using the method from Chapter 5. First make a common denominator of 24.
Multiply the numerator and denominator of $\frac{3}{8}$ by 3,
and multiply the numerator and denominator of $\frac{5}{6}$ by 4.
Note that $3 \times 3 + 5 \times 4 = 9 + 20 = 29$.)

227

❼ $\frac{7}{8}$ hours (Subtract the given fractions using the method from

Chapter 6. First make a common denominator of 8.

Multiply the numerator and denominator of $\frac{3}{2}$ by 4,

and leave $\frac{5}{8}$ unchanged. Note that $3 \times 4 - 5 = 12 - 5 = 7$.)

❽ $\frac{11}{60}$ hours (Subtract the given fractions using the method from

Chapter 6. First make a common denominator of 60.

Multiply the numerator and denominator of $\frac{7}{12}$ by 5,

and multiply the numerator and denominator of $\frac{2}{5}$ by 12.

Note that $7 \times 5 - 2 \times 12 = 35 - 24 = 11$.

The answer could also be expressed as 11 minutes,

since there are 60 minutes in an hour.)

❾ 5 (Multiply the given numbers using the method from Chapter 7.

Note that $30 = \frac{30}{1}$. Also note that $\frac{1}{6} \times \frac{30}{1} = \frac{30}{6} = 30 \div 6 = 5$.)

❿ 7 (First convert the percentage to a decimal. Divide by 100%,

or shift the decimal point two places to the left. You should get 0.25.

Now multiply the decimal and whole number using the method

from Chapter 14.)

$$
\begin{array}{r}
28 \\
\times\ 0.25 \\
\hline
7.00
\end{array}
$$

⓫ 15% (You should get $\frac{3}{20}$ as an intermediate answer.

Convert $\frac{3}{20}$ to a percentage using the method from Chapter 12.

Multiply the numerator and denominator of $\frac{3}{20}$ by 5.

You should get $\frac{15}{100}$, which equals 0.15. Multiply by 100%:

Shift the decimal point two places to the right to get 15%.

Enter $3 \div 20 \times 100$ on a calculator to verify that $\frac{3}{20} = 15\%$.)

⓬ the monkey (Compare $\frac{7}{4}$ and $\frac{8}{5}$ using a method from Chapter 4. If you use the cross multiplication method, you should get $35 > 32$. If you use the common denominator method, you should get $\frac{35}{20} > \frac{32}{20}$. Therefore, $\frac{7}{4} > \frac{8}{5}$, which shows that the monkey traveled farther.)

⓭ $1.22 (Add the decimals using the method from Chapter 13.)

$$\begin{array}{r} \$0.84 \\ + \$0.38 \\ \hline \$1.22 \end{array}$$

⓮ $1.75 (Subtract the decimals using the method from Chapter 13.)

$$\begin{array}{r} \$5.00 \\ - \$3.25 \\ \hline \$1.75 \end{array}$$

⓯ $2.25 (Multiply the decimals using the method from Chapter 14.)

$$\begin{array}{r} \$0.25 \\ \times\, 9 \\ \hline \$2.25 \end{array}$$

⓰ $0.30 (Divide $2.40 by 8 using the method from Chapter 14. Multiply 2.40 and 8 both by 100, and express them as a fraction. You should get $\frac{240}{800}$, which reduces to $\frac{3}{10}$ if you divide the numerator and denominator both by 80. Note that $\frac{3}{10}$ equals 0.3, which is the same as 0.30.)

❶⓱ $7.50 (First convert the percentage to a decimal. Divide by 100%, or shift the decimal point two places to the left. You should get 0.25. Now multiply the decimal and whole number using the method from Chapter 14.)

$$\begin{array}{r} \$30 \\ \times\ 0.25 \\ \hline \$7.50 \end{array}$$

❶⓲ $48.60 (First convert the percentage to a decimal. Divide by 100%, or shift the decimal point two places to the left. You should get 0.08. Now multiply the decimal and whole number using the method from Chapter 14. You should get $3.60. This is just the tax, not the answer. Now add the tax to the cost of the dress: $45 + $3.60. You should get $48.60.)

$$\begin{array}{r} \$45 \\ \times\ 0.08 \\ \hline \$3.60 \end{array} \qquad \begin{array}{r} \$45.00 \\ +\ \$3.60 \\ \hline \$48.60 \end{array}$$

❶⓳ $115.50 (First convert the percentages to decimals. Divide by 100%, or shift the decimal point two places to the left. You should get 0.3 and 0.1. Next multiply 0.3 × $150. You should get $45. Subtract $45 from $150 to determine the cost before tax. You should get $105.
Next multiply 0.1 × $105 to determine the tax. You should get $10.50. Now add $105 + $10.50. You should get $115.50.
Note: If you try to combine the 30% discount and 10% tax together to get 20%, this doesn't work because the discount and tax need to be applied to different amounts. Take the discount first, then apply the tax.)

$$\begin{array}{r} \$150 \\ \times\ 0.3 \\ \hline \$45 \end{array} \quad \begin{array}{r} \$150 \\ -\ \$45 \\ \hline \$105 \end{array} \quad \begin{array}{r} \$105 \\ \times\ 0.1 \\ \hline \$10.5 \end{array} \quad \begin{array}{r} \$105.00 \\ +\ \$10.50 \\ \hline \$115.50 \end{array}$$

❷⓴ $432 (First convert the percentage to a decimal. Divide by 100%, or shift the decimal point two places to the left. You should get 0.2. Now multiply the decimal and whole number. You should get $72. Add $360 + $72. You should get $432.)

Chapter 16 – Mixed Numbers

Part A

❶ $\frac{14}{3}$ (You should get $4 \times 3 + 2$ in the numerator.
Order of operations: Multiply first, then add.)

❷ $\frac{17}{2}$ (You should get $8 \times 2 + 1$ in the numerator.
If your answer is $\frac{23}{9}$, check the problem number carefully.)

❸ $\frac{19}{5}$ (You should get $3 \times 5 + 4$ in the numerator.
If your answer is $\frac{17}{2}$, check the problem number carefully.)

❹ $\frac{7}{4}$ (You should get $1 \times 4 + 3$ in the numerator.
If your answer is $\frac{45}{7}$, check the problem number carefully.)

❺ $\frac{23}{9}$ (You should get $2 \times 9 + 5$ in the numerator.
If your answer is $\frac{19}{5}$, check the problem number carefully.)

❻ $\frac{45}{7}$ (You should get $6 \times 7 + 3$ in the numerator.
If your answer is $\frac{55}{6}$, check the problem number carefully.)

❼ $\frac{55}{6}$ (You should get $9 \times 6 + 1$ in the numerator.
If your answer is $\frac{7}{4}$, check the problem number carefully.)

❽ $\frac{47}{8}$ (You should get $5 \times 8 + 7$ in the numerator.)

Part B

Note: You can check your answers in Part B by converting your answer back into an improper fraction. For example, for Problem ❾, if you convert the answer, $2\frac{2}{3}$, into an improper fraction, you get $2 \times 3 + 2 = 6 + 2 = 8$, which leads to $\frac{8}{3}$.

❾ $2\frac{2}{3}$ (You should get $8 \div 3 = 2$ R2, where 2 R2 means 2 with a remainder of 2. Note that $2 \times 3 = 6$ and $8 - 6 = 2$.)

❿ $4\frac{2}{5}$ (You should get $22 \div 5 = 4$ R2. Note that $4 \times 5 = 20$ and $22 - 20 = 2$. If your answer is $3\frac{4}{9}$, check the problem number carefully.)

⓫ $4\frac{5}{6}$ (You should get $29 \div 6 = 4$ R5. Note that $4 \times 6 = 24$ and $29 - 24 = 5$. If your answer is $4\frac{2}{5}$, check the problem number carefully.)

⓬ $5\frac{1}{2}$ (You should get $11 \div 2 = 5$ R1. Note that $5 \times 2 = 10$ and $11 - 10 = 1$. If your answer is $6\frac{3}{4}$, check the problem number carefully.)

⓭ $3\frac{4}{9}$ (You should get $31 \div 9 = 3$ R4. Note that $3 \times 9 = 27$ and $31 - 27 = 4$. If your answer is $4\frac{5}{6}$, check the problem number carefully.)

⓮ $6\frac{3}{4}$ (You should get $27 \div 4 = 6$ R3. Note that $6 \times 4 = 24$ and $27 - 24 = 3$. If your answer is $1\frac{2}{7}$, check the problem number carefully.)

⓯ $1\frac{2}{7}$ (You should get $9 \div 7 = 1$ R2. Note that $1 \times 7 = 7$ and $9 - 7 = 2$. If your answer is $5\frac{1}{2}$, check the problem number carefully.)

⓰ $6\frac{5}{8}$ (You should get $53 \div 8 = 6$ R5, where 6 R5 means 6 with a remainder of 5. Note that $6 \times 8 = 48$ and $53 - 48 = 5$.)

Chapter 17 – Reciprocals of Mixed Numbers

Part A

❶ $\frac{2}{7}$ (You should get $3 \times 2 + 1$ in the original numerator.

Then swap the numerator and denominator to find the reciprocal.)

❷ $\frac{8}{37}$ (You should get $4 \times 8 + 5$ in the original numerator.

If your answer is $\frac{5}{29}$, check the problem number carefully.)

❸ $\frac{5}{9}$ (You should get $1 \times 5 + 4$ in the original numerator.

If your answer is $\frac{8}{37}$, check the problem number carefully.)

❹ $\frac{4}{31}$ (You should get $7 \times 4 + 3$ in the original numerator.

If your answer is $\frac{7}{18}$, check the problem number carefully.)

❺ $\frac{5}{29}$ (You should get $5 \times 5 + 4$ in the original numerator.

If your answer is $\frac{5}{9}$, check the problem number carefully.)

❻ $\frac{7}{18}$ (You should get $2 \times 7 + 4$ in the original numerator.

If your answer is $\frac{5}{42}$, check the problem number carefully.)

❼ $\frac{5}{42}$ (You should get $8 \times 5 + 2$ in the original numerator.

If your answer is $\frac{4}{31}$, check the problem number carefully.)

❽ $\frac{9}{89}$ (You should get $9 \times 9 + 8$ in the original numerator.)

Part B

❾ $\frac{8}{47}$ (You should get $5 \times 8 + 7$ in the original numerator.

Then swap the numerator and denominator to find the reciprocal)

❿ $\frac{9}{55}$ (You should get $6 \times 9 + 1$ in the original numerator.

If your answer is $\frac{3}{29}$, check the problem number carefully.)

⓫ $\frac{9}{20}$ (You should get $2 \times 9 + 2$ in the original numerator.

If your answer is $\frac{9}{55}$, check the problem number carefully.)

⓬ $\frac{2}{3}$ (You should get $1 \times 2 + 1$ in the original numerator.

If your answer is $\frac{9}{41}$, check the problem number carefully.)

⓭ $\frac{3}{29}$ (You should get $9 \times 3 + 2$ in the original numerator.

If your answer is $\frac{9}{20}$, check the problem number carefully.)

⓮ $\frac{9}{41}$ (You should get $4 \times 9 + 5$ in the original numerator.

If your answer is $\frac{3}{10}$, check the problem number carefully.)

⓯ $\frac{3}{10}$ (You should get $3 \times 3 + 1$ in the original numerator.

If your answer is $\frac{2}{3}$, check the problem number carefully.)

⓰ $\frac{7}{59}$ (You should get $8 \times 7 + 3$ in the original numerator.)

Chapter 18 – Arithmetic with Mixed Numbers

Part A

❶ $3\frac{14}{15}$ (Note that $1 \times 5 + 3 = 8$ and $2 \times 3 + 1 = 7$.

The given problem is equivalent to $\frac{8}{5} + \frac{7}{3}$.

Multiply $\frac{8}{5}$ by $\frac{3}{3}$ and multiply $\frac{7}{3}$ by $\frac{5}{5}$ to make a common denominator.

You should get $\frac{24}{15} + \frac{35}{15} = \frac{59}{15}$. Convert the improper fraction $\frac{59}{15}$ into a mixed number using the method from Chapter 16.

Note that $45 \div 15 = 3$ and $59 - 45 = 14$ such that $\frac{59}{15} = 3\frac{14}{15}$.

Calculator check: $1 + \frac{3}{5} + 2 + \frac{1}{3}$ and $3 + \frac{14}{15}$ both equal $3.9\bar{3}$.)

❷ $8\frac{1}{6}$ (Note that $3 \times 2 + 1 = 7$ and $4 \times 3 + 2 = 14$.

The given problem is equivalent to $\frac{7}{2} + \frac{14}{3}$.

Multiply $\frac{7}{2}$ by $\frac{3}{3}$ and multiply $\frac{14}{3}$ by $\frac{2}{2}$ to make a common denominator.

You should get $\frac{21}{6} + \frac{28}{6} = \frac{49}{6}$. Convert the improper fraction $\frac{49}{6}$ into a mixed number using the method from Chapter 16.

Note that $48 \div 6 = 8$ and $49 - 48 = 1$ such that $\frac{49}{6} = 8\frac{1}{6}$.

Calculator check: $3 + \frac{1}{2} + 4 + \frac{2}{3}$ and $8 + \frac{1}{6}$ both equal $8.1\bar{6}$.)

❸ $7\frac{11}{12}$ (Note that $5 \times 4 + 3 = 23$ and $2 \times 6 + 1 = 13$.

The given problem is equivalent to $\frac{23}{4} + \frac{13}{6}$.

Multiply $\frac{23}{4}$ by $\frac{3}{3}$ and multiply $\frac{13}{6}$ by $\frac{2}{2}$ to make a common denominator.

You should get $\frac{69}{12} + \frac{26}{12} = \frac{95}{12}$. Convert the improper fraction $\frac{95}{12}$ into a mixed number using the method from Chapter 16.

Note that $84 \div 12 = 7$ and $95 - 84 = 11$ such that $\frac{95}{12} = 7\frac{11}{12}$.

Calculator check: $5 + \frac{3}{4} + 2 + \frac{1}{6}$ and $7 + \frac{14}{15}$ both equal $7.91\bar{6}$.)

❹ 16 (Note that $7 \times 6 + 1 = 43$ and $8 \times 6 + 5 = 53$.

The given problem is equivalent to $\frac{43}{6} + \frac{53}{6}$.

The denominator is already common. Just add the numerators.

You should get $\frac{43}{6} + \frac{53}{6} = \frac{96}{6}$. Note that $\frac{96}{6} = 96 \div 6 = 16$.

Calculator check: $7 + \frac{1}{6} + 8 + \frac{5}{6}$ equals 16.)

❺ $11\frac{3}{10}$ (Note that $6 \times 5 + 4 = 34$ and $4 \times 2 + 1 = 9$.

The given problem is equivalent to $\frac{34}{5} + \frac{9}{2}$.

Multiply $\frac{34}{5}$ by $\frac{2}{2}$ and multiply $\frac{9}{2}$ by $\frac{5}{5}$ to make a common denominator.

You should get $\frac{68}{10} + \frac{45}{10} = \frac{113}{10}$. Convert the improper fraction $\frac{113}{10}$

into a mixed number using the method from Chapter 16.

Note that $110 \div 10 = 11$ and $113 - 110 = 3$ such that $\frac{113}{10} = 11\frac{3}{10}$.

Calculator check: $6 + \frac{4}{5} + 4 + \frac{1}{2}$ and $11 + \frac{3}{10}$ both equal 11.3.)

Part B

❻ $2\frac{5}{6}$ (Note that $5 \times 3 + 1 = 16$ and $2 \times 2 + 1 = 5$.

The given problem is equivalent to $\frac{16}{3} - \frac{5}{2}$.

Multiply $\frac{16}{3}$ by $\frac{2}{2}$ and multiply $\frac{5}{2}$ by $\frac{3}{3}$ to make a common denominator.

You should get $\frac{32}{6} - \frac{15}{6} = \frac{17}{6}$. Convert the improper fraction $\frac{17}{6}$ into a mixed number using the method from Chapter 16.

Note that $12 \div 6 = 2$ and $17 - 12 = 5$ such that $\frac{17}{6} = 2\frac{5}{6}$.

Calculator check: $5 + \frac{1}{3} - 2 - \frac{1}{2}$ and $2 + \frac{5}{6}$ both equal $2.8\overline{3}$.

Note the two minus signs in the calculator check. This has to do with the distributive property.)

❼ $2\frac{17}{20}$ (Note that $6 \times 4 + 1 = 25$ and $3 \times 5 + 2 = 17$.

The given problem is equivalent to $\frac{25}{4} - \frac{17}{5}$.

Multiply $\frac{25}{4}$ by $\frac{5}{5}$ and multiply $\frac{17}{5}$ by $\frac{4}{4}$ to make a common denominator.

You should get $\frac{125}{20} - \frac{68}{20} = \frac{57}{20}$. Convert the improper fraction $\frac{57}{20}$ into a mixed number using the method from Chapter 16.

Note that $40 \div 20 = 2$ and $57 - 40 = 17$ such that $\frac{57}{20} = 2\frac{17}{20}$.

Calculator check: $6 + \frac{1}{4} - 3 - \frac{2}{5}$ and $2 + \frac{17}{20}$ both equal 2.85.)

❽ $3\frac{11}{24}$ (Note that $4 \times 8 + 5 = 37$ and $1 \times 6 + 1 = 7$.

The given problem is equivalent to $\frac{37}{8} - \frac{7}{6}$.

Multiply $\frac{37}{8}$ by $\frac{3}{3}$ and multiply $\frac{7}{6}$ by $\frac{4}{4}$ to make a common denominator.

You should get $\frac{111}{24} - \frac{28}{24} = \frac{83}{24}$. Convert the improper fraction $\frac{83}{24}$ into a mixed number using the method from Chapter 16.

Note that $72 \div 24 = 3$ and $83 - 72 = 11$ such that $\frac{83}{24} = 3\frac{11}{24}$.

Calculator check: $4 + \frac{5}{8} - 1 - \frac{1}{6}$ and $3 + \frac{11}{24}$ both equal $3.458\overline{3}$.)

❾ $3\frac{1}{6}$ (Note that $4 \times 6 + 5 = 29$. The given problem is equivalent to $\frac{8}{1} - \frac{29}{6}$.

Note that $8 = \frac{8}{1}$. Multiply $\frac{8}{1}$ by $\frac{6}{6}$ to make a common denominator.

You should get $\frac{48}{6} - \frac{29}{6} = \frac{19}{6}$. Convert the improper fraction $\frac{19}{6}$

into a mixed number using the method from Chapter 16.

Note that $18 \div 6 = 3$ and $19 - 18 = 1$ such that $\frac{19}{6} = 3\frac{1}{6}$.

Calculator check: $8 - 4 - \frac{5}{6}$ and $3 + \frac{1}{6}$ both equal $3.1\overline{6}$.

Note the two minus signs in the calculator check. This has to do with the distributive property.)

❿ $2\frac{5}{8}$ (Note that $9 \times 8 + 5 = 77$. The given problem is equivalent to $\frac{77}{8} - \frac{7}{1}$.

Note that $7 = \frac{7}{1}$. Multiply $\frac{7}{1}$ by $\frac{8}{8}$ to make a common denominator.

You should get $\frac{77}{8} - \frac{56}{8} = \frac{21}{8}$. Convert the improper fraction $\frac{21}{8}$

into a mixed number using the method from Chapter 16.

Note that $16 \div 8 = 2$ and $21 - 16 = 5$ such that $\frac{21}{8} = 2\frac{5}{8}$.

Calculator check: $9 + \frac{5}{8} - 7$ and $2 + \frac{5}{8}$ both equal 2.625.

If you think about this problem, you may realize that there is a much shorter way to solve it. However, if you figure out the short solution, spend a minute thinking about why the short solution doesn't work on Exercise 9. Students who are aware of the short solution to Exercise 10 often make mistakes when solving problems like Exercise 9.)

Part C

⓫ 12 (Note that $4 \times 2 + 1 = 9$ and $2 \times 3 + 2 = 8$.

The given problem is equivalent to $\frac{9}{2} \times \frac{8}{3}$. Note that $\frac{72}{6} = 72 \div 6 = 12$

because $6 \times 12 = 72$. The final answer is 12.

Calculator check: $\left(4 + \frac{1}{2}\right) \times \left(2 + \frac{2}{3}\right)$ equals 12.)

⓬ 28 (Note that $6 \times 3 + 2 = 20$ and $4 \times 5 + 1 = 21$.

The given problem is equivalent to $\frac{20}{3} \times \frac{21}{5}$. Note that $\frac{420}{15} = 420 \div 15 = 28$

because $15 \times 28 = 420$. The final answer is 28.

Calculator check: $\left(6 + \frac{2}{3}\right) \times \left(4 + \frac{1}{5}\right)$ equals 28.)

⓭ $29\frac{1}{6}$ (Note that $8 \times 4 + 3 = 35$ and $3 \times 3 + 1 = 10$.

The given problem is equivalent to $\frac{35}{4} \times \frac{10}{3}$. Note that $\frac{350}{12}$ reduces to $\frac{175}{6}$.

Note that $174 \div 6 = 29$ since $6 \times 29 = 174$.

Also note that $175 - 174 = 1$ such that $\frac{175}{6} = 29\frac{1}{6}$.

Calculator check: $\left(8 + \frac{3}{4}\right) \times \left(3 + \frac{1}{3}\right)$ and $29 + \frac{1}{6}$ both equal $29.1\bar{6}$.)

⓮ $33\frac{1}{3}$ (Note that $6 \times 3 + 2 = 20$. The given problem is equivalent to $\frac{20}{3} \times \frac{5}{1}$.

Note that $5 = \frac{5}{1}$. Note that $100 \div 3 = 33$ and $100 - 99 = 1$ such that

$\frac{100}{3} = 33\frac{1}{3}$. Calculator check: $\left(6 + \frac{2}{3}\right) \times 5$ and $33 + \frac{1}{3}$ both equal $33.\bar{3}$.)

⓯ $4\frac{3}{8}$ (Note that $2 \times 2 + 1 = 5$ and $1 \times 4 + 3 = 7$.

The given problem is equivalent to $\frac{5}{2} \times \frac{7}{4}$. You should get $\frac{35}{8}$.

Note that $32 \div 8 = 4$ and $35 - 32 = 3$ such that $\frac{35}{8} = 4\frac{3}{8}$.

Calculator check: $\left(2 + \frac{1}{2}\right) \times \left(1 + \frac{3}{4}\right)$ and $4 + \frac{3}{8}$ both equal 4.375.)

Part D

⑯ $2\frac{1}{12}$ (Note that $6 \times 3 + 2 = 20$ and $3 \times 5 + 1 = 16$.

To divide by a fraction, multiply by its reciprocal: $\frac{20}{3} \div \frac{16}{5} = \frac{20}{3} \times \frac{5}{16}$.

Note that $\frac{100}{48}$ reduces to $\frac{25}{12}$. Note that $24 \div 12 = 2$ and $25 - 24 = 1$ such that

$\frac{25}{12} = 2\frac{1}{12}$. Calculator check: $\left(6 + \frac{2}{3}\right) \div \left(3 + \frac{1}{5}\right)$ and $2 + \frac{1}{12}$ both equal $2.08\overline{3}$.)

⑰ $1\frac{59}{75}$ (Note that $7 \times 9 + 4 = 67$ and $4 \times 6 + 1 = 25$.

To divide by a fraction, multiply by its reciprocal: $\frac{67}{9} \div \frac{25}{6} = \frac{67}{9} \times \frac{6}{25}$.

Note that $\frac{402}{225}$ reduces to $\frac{134}{75}$. Note that $402 \div 3 = 134$ and $225 \div 3 = 75$.

Note that $75 \div 75 = 1$ and $134 - 75 = 59$ such that $\frac{134}{75} = 1\frac{59}{75}$.

Calculator check: $\left(7 + \frac{4}{9}\right) \div \left(4 + \frac{1}{6}\right)$ and $1 + \frac{59}{75}$ both equal $1.78\overline{6}$.)

⑱ $7\frac{1}{2}$ (Note that $3 \times 4 + 3 = 15$. To divide by a fraction, multiply by its

reciprocal: $\frac{15}{4} \div \frac{1}{2} = \frac{15}{4} \times \frac{2}{1}$. Note that $\frac{30}{4}$ reduces to $\frac{15}{2}$.

Note that $14 \div 2 = 7$ and $15 - 14 = 1$ such that $\frac{15}{2} = 7\frac{1}{2}$.

Calculator check: $\left(3 + \frac{3}{4}\right) \div \frac{1}{2}$ and $7 + \frac{1}{2}$ both equal 7.5.)

⑲ $2\frac{2}{7}$ (Note that $2 \times 8 + 5 = 21$. To divide by a fraction, multiply by its

reciprocal: $\frac{6}{1} \div \frac{21}{8} = \frac{6}{1} \times \frac{8}{21}$. Note that $\frac{48}{21}$ reduces to $\frac{16}{7}$.

Note that $14 \div 7 = 2$ and $16 - 14 = 2$ such that $\frac{16}{7} = 2\frac{2}{7}$.

Calculator check: $6 \div \left(2 + \frac{5}{8}\right)$ and $2 + \frac{2}{7}$ both equal $2.\overline{285714}$.)

⑳ $2\frac{22}{27}$ (Note that $8 \times 9 + 4 = 76$. Note that the reciprocal of 3 is $\frac{1}{3}$.

To divide by a fraction, multiply by its reciprocal: $\frac{76}{9} \div \frac{3}{1} = \frac{76}{9} \times \frac{1}{3} = \frac{76}{27}$.

Note that $54 \div 27 = 2$ and $76 - 54 = 22$ such that $\frac{76}{27} = 2\frac{22}{27}$.

Calculator check: $\left(8 + \frac{4}{9}\right) \div 3$ and $2 + \frac{22}{27}$ both equal $2.\overline{814}$.)

Chapter 19 – Ratios and Proportions

Part A

❶ 36 (Multiply $\frac{12}{8}$ by $\frac{3}{3}$.)

❷ 9 (Multiply $\frac{3}{5}$ by $\frac{3}{3}$.
If your answer is 16, check the problem number carefully.)

❸ 5 (First reduce $\frac{3}{6}$ to $\frac{1}{2}$ by dividing the numerator and denominator
both by 3, and then multiply $\frac{1}{2}$ by $\frac{5}{5}$.
If your answer is 9, check the problem number carefully.)

❹ 4 (Reduce $\frac{24}{18}$ by dividing the numerator and denominator by 6.
If your answer is 12, check the problem number carefully.)

❺ 16 (Multiply $\frac{4}{7}$ by $\frac{4}{4}$.
If your answer is 5, check the problem number carefully.)

❻ 12 (First reduce $\frac{9}{3}$ to $\frac{3}{1}$ by dividing the numerator and denominator
both by 3, and then multiply $\frac{3}{1}$ by $\frac{4}{4}$.
If your answer is 7, check the problem number carefully.)

❼ 7 (Reduce $\frac{28}{44}$ by dividing the numerator and denominator by 4.
If your answer is 4, check the problem number carefully.)

❽ 6 (First reduce $\frac{15}{10}$ to $\frac{3}{2}$ by dividing the numerator and denominator
both by 5, and then multiply $\frac{3}{2}$ by $\frac{2}{2}$.)

Part B

❾ 81 (Multiply $\frac{9}{8}$ by $\frac{9}{9}$.)

❿ 36 (First reduce $\frac{45}{25}$ to $\frac{9}{5}$ by dividing the numerator and denominator both by 5, and then multiply $\frac{9}{5}$ by $\frac{4}{4}$.
If your answer is 1, check the problem number carefully.)

⓫ 24 (Multiply $\frac{3}{2}$ by $\frac{8}{8}$.
If your answer is 36, check the problem number carefully.)

⓬ 10 (First reduce $\frac{35}{28}$ to $\frac{5}{4}$ by dividing the numerator and denominator both by 7, and then multiply $\frac{5}{4}$ by $\frac{2}{2}$.
If your answer is 32, check the problem number carefully.)

⓭ 1 (Reduce $\frac{5}{100}$ by dividing the numerator and denominator by 5.
If your answer is 24, check the problem number carefully.)

⓮ 32 (First reduce $\frac{24}{21}$ to $\frac{8}{7}$ by dividing the numerator and denominator both by 3, and then multiply $\frac{8}{7}$ by $\frac{4}{4}$.
If your answer is 95, check the problem number carefully.)

⓯ 95 (First reduce $\frac{13}{13}$ to $\frac{1}{1}$ by dividing the numerator and denominator both by 13, and then multiply $\frac{1}{1}$ by $\frac{95}{95}$.
Alternatively, just realize that the numerator equals the denominator.
If your answer is 10, check the problem number carefully.)

⓰ 15 (First reduce $\frac{6}{16}$ to $\frac{3}{8}$ by dividing the numerator and denominator both by 2, and then multiply $\frac{3}{8}$ by $\frac{5}{5}$.)

Chapter 20 – Long Division

Part A

❶

$$
\begin{array}{r}
0.375 \\
8\overline{)3.0} \\
-2.4 \\
\hline
0.60 \\
-0.56 \\
\hline
0.040 \\
-0.040 \\
\hline
0.000
\end{array}
$$

$8 \times 0.3 = 2.4$

$8 \times 0.07 = 0.56$

$8 \times 0.005 = 0.040$

$$\frac{3}{8} = 0.375$$

❷

$$
\begin{array}{r}
0.454545 \\
11\overline{)5.0} \\
-4.4 \\
\hline
0.60 \\
-0.55 \\
\hline
0.050 \\
-0.044 \\
\hline
0.0060 \\
-0.0055 \\
\hline
0.00050
\end{array}
$$

$11 \times 0.4 = 4.4$

$11 \times 0.05 = 0.55$

$11 \times 0.004 = 0.044$

$11 \times 0.0005 = 0.0055$

$$\frac{5}{11} = 0.\overline{45} \text{ (note the bar)}$$

Part B

❸

$$
\begin{array}{r}
1.1666 \\
6\overline{)7} \\
-6 \\
\hline
1.0 \\
-0.6 \\
\hline
0.40 \\
-0.36 \\
\hline
0.040 \\
-0.036 \\
\hline
0.0040
\end{array}
$$

$6 \times 1 = 6$

$6 \times 0.1 = 0.6$

$6 \times 0.06 = 0.36$

$6 \times 0.006 = 0.036$

$$\frac{7}{6} = 1.1\overline{6} \text{ (note the bar)}$$

❹

$$
\begin{array}{r}
0.4444 \\
9\overline{)4.0} \\
-3.6 \\
\hline
0.40 \\
-0.36 \\
\hline
0.040 \\
-0.036 \\
\hline
0.0040
\end{array}
$$

$9 \times 0.4 = 3.6$

$9 \times 0.04 = 0.36$

$9 \times 0.004 = 0.036$

$$\frac{4}{9} = 0.\overline{4} \text{ (note the bar)}$$

Part C

❺

$$
\begin{array}{r}
0.4375 \\
16\ \overline{)7.0} \\
\end{array}
$$

-6.4 $16 \times 0.4 = 6.4$
$$\overline{0.60}$$
-0.48 $16 \times 0.03 = 0.48$
$$\overline{0.120}$$
-0.112 $16 \times 0.007 = 0.112$
$$\overline{0.008}$$
-0.008 $16 \times 0.0005 = 0.0080$
$$\overline{0.000}$$

$$\frac{7}{16} = 0.4375$$

❻

$$
\begin{array}{r}
0.213333 \\
75\ \overline{)16.0} \\
\end{array}
$$

-15.0 $75 \times 0.2 = 15.0$
$$\overline{1.00}$$
$-\ 0.75$ $75 \times 0.01 = 0.75$
$$\overline{0.250}$$
$-\ 0.225$ $75 \times 0.003 = 0.225$
$$\overline{0.0250}$$
$-\ 0.0225$ $75 \times 0.0003 = 0.0225$
$$\overline{0.00250}$$

$$\frac{16}{75} = 0.21\overline{3} \text{ (note the bar)}$$

Part D

❼

$$
\begin{array}{r}
0.425 \\
40\ \overline{)17.0} \\
\end{array}
$$

-16.0 $40 \times 0.4 = 16.0$
$$\overline{1.00}$$
$-\ 0.80$ $40 \times 0.02 = 0.80$
$$\overline{0.200}$$
$-\ 0.200$ $40 \times 0.005 = 0.200$
$$\overline{0.000}$$

$$\frac{17}{40} = 0.425$$

❽

$$
\begin{array}{r}
0.41666 \\
12\ \overline{)5.0} \\
\end{array}
$$

-4.8 $12 \times 0.4 = 4.8$
$$\overline{0.20}$$
-0.12 $12 \times 0.01 = 0.12$
$$\overline{0.080}$$
-0.072 $12 \times 0.006 = 0.072$
$$\overline{0.0080}$$
-0.0072 $12 \times 0.0006 = 0.0072$
$$\overline{0.00080}$$

$$\frac{5}{12} = 0.41\overline{6} \text{ (note the bar)}$$

Posttest Assessment

❶ $\frac{3}{8}$ (Divide the numerator and denominator both by 5.
Note that $15 \div 5 = 3$ and $40 \div 5 = 8$. See Chapter 2.)

❷ $\frac{2}{3}$ (One method is to make a common denominator:
Multiply the numerator and denominator of $\frac{2}{3}$ by 8,
and multiply the numerator and denominator of $\frac{5}{8}$ by 3.
You should get $\frac{2}{3} = \frac{16}{24}$ and $\frac{5}{8} = \frac{15}{24}$. Since $16 > 15$, it follows that $\frac{2}{3} > \frac{5}{8}$.
Alternatively, cross multiply to get $2 \times 8 > 5 \times 3$ or $16 > 15$.
See Chapter 4.)

❸ $\frac{7}{4}$ (Swap the numerator and denominator of $\frac{4}{7}$. See Chapter 8.)

❹ $\frac{1}{5}$ (First write $5 = \frac{5}{1}$. Then swap the numerator and denominator of $\frac{5}{1}$.
See Chapter 8.)

❺ $\frac{11}{12}$ (Find a common denominator. Multiply the numerator
and denominator of $\frac{3}{4}$ by 3, and multiply the numerator and denominator
of $\frac{1}{6}$ by 2. You should get $\frac{9}{12} + \frac{2}{12} = \frac{11}{12}$. See Chapter 5.)

❻ $\frac{17}{45}$ (Find a common denominator. Multiply the numerator
and denominator of $\frac{3}{5}$ by 9, and multiply the numerator and denominator
of $\frac{2}{9}$ by 5. You should get $\frac{27}{45} - \frac{10}{45} = \frac{17}{45}$. See Chapter 6.)

❼ $\frac{5}{6}$ (Multiply the numerators to get the new numerator: $2 \times 5 = 10$.
Multiply the denominators to get the new denominator: $3 \times 4 = 12$.
Reduce $\frac{10}{12}$: Divide the numerator and denominator both by 2, using
the method from Chapter 2: $\frac{10}{12} = \frac{10 \div 2}{12 \div 2} = \frac{5}{6}$. See Chapter 7.)

❽ $\frac{5}{9}$ (Dividing by a fraction is the same as multiplying by the reciprocal of the second fraction. The reciprocal of $\frac{3}{4}$ is $\frac{4}{3}$, as discussed in Chapter 8. Therefore, $\frac{5}{12} \div \frac{3}{4} = \frac{5}{12} \times \frac{4}{3}$. Multiply $\frac{5}{12}$ and $\frac{4}{3}$.

Multiply the numerators to get the new numerator: $5 \times 4 = 20$.

Multiply the denominators to get the new denominator: $12 \times 3 = 36$.

Reduce $\frac{20}{36}$: Divide the numerator and denominator both by 4, using

the method from Chapter 2: $\frac{20}{36} = \frac{20 \div 4}{36 \div 4} = \frac{5}{9}$. See Chapter 9.

If you get $\frac{10}{18}$, you didn't fully reduce your answer.)

❾ $\frac{3}{4}$ (Since the last digit of 0.75, which is a 5, is in the hundredths place,

0.75 means $\frac{75}{100}$. Reduce your answer using the method from Chapter 2.

Divide the numerator and denominator both by 25. Note that $75 \div 25 = 3$

and $100 \div 25 = 4$, such that $\frac{75}{100} = \frac{3}{4}$. See Chapter 10.)

❿ 0.625 (One method is to multiply both the numerator and denominator

of $\frac{5}{8}$ by 125. Note that $5 \times 125 = 625$ and $8 \times 125 = 1000$, such that $\frac{5}{8} = \frac{625}{1000}$.

Since 625 is over 1000, express the numerator as a decimal with the last

digit, which is a 5, in the thousandths place: $\frac{625}{1000} = 0.625$. See Chapter 10.

An alternate method is to use long division, as shown in Chapter 20.)

$$
\begin{array}{r}
0.625 \\
8\overline{)5.0} \\
\end{array}
$$

-4.8 $8 \times 0.6 = 4.8$

0.20

-0.16 $8 \times 0.02 = 0.16$

0.040

-0.040 $8 \times 0.005 = 0.040$

0.000

$$\frac{5}{8} = 0.625$$

⓫ 40% (Multiply 0.4 by 100%. This effectively shifts the decimal point two places to the right, changing 0.4 into 40%. See Chapter 12.)

⑫ 250% (One method is to multiply the numerator and denominator of $\frac{5}{2}$ by 50. You should get $\frac{250}{100}$. Since 250 is over 100, express the numerator as a decimal with the last digit, which is a 0, in the hundredths place: $\frac{250}{100} = 2.50$. To convert from a decimal to a percentage, multiply by 100% or shift the decimal point two places to the right: $2.50 = 250\%$. Enter $5 \div 2 \times 100$ on a calculator to verify that $\frac{5}{2} = 250\%$. See Chapter 12. Alternatively, first multiply $\frac{5}{2}$ by 100% to get $\frac{500}{2}\% = 500\% \div 2 = 250\%$.)

⑬ $\frac{7}{20}$ (Since 35% means 35 out of 100, make a fraction with 35 in the numerator and 100 in the denominator: $35\% = \frac{35}{100}$. Reduce your answer using the method from Chapter 2. Note that $35 \div 5 = 7$ and $100 \div 5 = 20$, such that $\frac{35}{100} = \frac{7}{20}$. See Chapter 12.)

⑭ 1.5 (Divide 150 by 100%. This effectively shifts the decimal point two places to the left, changing 150% into 1.5. See Chapter 12.)

⑮ $2.\overline{6}$ (One method is to multiply the numerator and denominator of $\frac{8}{3}$ by 3 to get $\frac{24}{9}$. Since the numerator is larger than the denominator, rewrite $\frac{24}{9}$ as a whole number plus a fraction that is smaller than one. Note that $24 \div 9$ equals 2 with a remainder of 6 (since $9 \times 2 = 18$ and $24 - 18 = 6$). Thus, $\frac{24}{9} = 2 + \frac{6}{9}$. That's the whole number plus the remainder divided by 9. Basically, we just converted an improper fraction into a mixed number using the method of Chapter 16. Note that $\frac{6}{9} = 0.\overline{6}$ according to Chapter 11. Plug $\frac{6}{9} = 0.\overline{6}$ into $\frac{24}{9} = 2 + \frac{6}{9}$ to get $\frac{24}{9} = 2 + 0.\overline{6} = 2.\overline{6}$.
Look closely: 2.6 is incorrect. Put a bar over the 6 in $2.\overline{6}$.
This means 2.666666... where the 6 repeats forever.
You can enter $8 \div 3$ on a calculator to check that $\frac{8}{3} = 2.66666666...$
Note that a calculator will round the last 6 up to a 7. See Chapter 11.
An alternate method is to use long division, as shown in Chapter 20.)

$$\begin{array}{r} 2.666 \\ 3\overline{\smash{)}8} \\ -6 \qquad 3 \times 2 = 6 \\ \hline 2.0 \\ -1.8 \qquad 3 \times 0.6 = 1.8 \\ \hline 0.20 \\ 8 \end{array}$$

$$\frac{8}{3} = 2.\overline{6} \text{ (note the bar)}$$

❶❻ $\frac{5}{11}$ (Since two digits are repeating, divide 45 by 99 to get $\frac{45}{99}$. Reduce your answer using the method from Chapter 2. Note that $45 \div 9 = 5$ and $99 \div 9 = 11$, such that $\frac{45}{99} = \frac{5}{11}$. You can enter $5 \div 11$ on a calculator to check that $\frac{5}{11} = 0.45454545\ldots$ Incorrect answers include $\frac{9}{20}$ and $\frac{45}{100}$. If you get these, note that $0.\overline{45}$ means 0.45454545 with the 45 repeating forever. Note that $0.\overline{45}$ doesn't mean the same thing as 0.45. See Chapter 11.)

❶❼ $6\frac{2}{3}$ (Note that $\frac{20}{3} = 20 \div 3 = 6$ R2 meaning 6 with a remainder of 2. Note that $18 \div 3 = 6$ and $20 - 18 = 2$. See Chapter 16.)

❶❽ $\frac{9}{4}$ (The new numerator is $2 \times 4 + 1 = 9$. See Chapter 16.)

❶❾ $9\frac{3}{20}$ (Note that $3 \times 5 + 2 = 17$ and $5 \times 4 + 3 = 23$, such that $3\frac{2}{5} = \frac{17}{5}$ and $5\frac{3}{4} = \frac{23}{4}$. Multiply $\frac{17}{5}$ by $\frac{4}{4}$ and multiply $\frac{23}{3}$ by $\frac{5}{5}$ to make a common denominator. You should get $\frac{68}{20} + \frac{115}{20} = \frac{183}{20}$. Convert the improper fraction $\frac{183}{20}$ into a mixed number using the method of Chapter 16. Note that $180 \div 20 = 9$ and $183 - 180 = 3$ such that $\frac{183}{20} = 9\frac{3}{20}$. Calculator check: $3 + \frac{2}{5} + 5 + \frac{3}{4}$ and $9 + \frac{3}{20}$ both equal 9.15. If you get $8\frac{23}{20}$, your answer is correct, but your answer isn't in standard form. Note that $8\frac{23}{20} = 9\frac{3}{20}$ because $\frac{23}{20} = 1\frac{3}{20}$.)

❷⓿ $\frac{2}{3}$ (Add the given numbers together to find the denominator: $6 + 12 = 18$. You should get $\frac{12}{18}$ as an intermediate answer. Reduce your answer using the method from Chapter 2. Note that $\frac{12}{18} = \frac{12 \div 6}{18 \div 6} = \frac{2}{3}$. See Chapter 15.)

WAS THIS BOOK HELPFUL?

A great deal of effort and thought was put into this book, such as:

- Breaking down the solutions to help make the math easier to understand.
- Careful selection of examples and problems for their instructional value.
- Thorough answer key including several explanations.
- Preparing a pretest/posttest to help measure learning improvement.
- Multiple stages of proofreading, editing, and formatting.
- Beta testers provided valuable feedback.

If you appreciate the effort that went into making this book possible, there is a simple way that you could show it:

Please take a moment to post an honest review.

For example, you can review this book at Amazon.com or Barnes & Noble's website at BN.com.

Even a short review can be helpful and will be much appreciated. If you're not sure what to write, following are a few ideas, though it's best to describe what's important to you.

- Were you able to understand the explanations?
- How much did you learn from reading and using this workbook?
- Was it helpful to follow the examples while solving the problems?
- Did the answer key help you understand the solutions?
- Would you recommend this book to others? If so, why?

Do you believe that you found a mistake? Please email the author, Chris McMullen, at greekphysics@yahoo.com to ask about it. One of two things will happen:

- You might discover that it wasn't a mistake after all and learn why.
- You might be right, in which case the author will be grateful and future readers will benefit from the correction. Everyone is human.

ABOUT THE AUTHOR

Dr. Chris McMullen has over 20 years of experience teaching university physics in California, Oklahoma, Pennsylvania, and Louisiana. Dr. McMullen is also an author of math and science books. Whether in the classroom or as a writer, Dr. McMullen loves sharing knowledge and the art of motivating and engaging students.

Chris McMullen earned his Ph.D. in phenomenological high-energy physics (particle physics) from Oklahoma State University in 2002. Originally from California, Dr. McMullen earned his Master's degree from California State University, Northridge, where his thesis was in the field of electron spin resonance.

As a physics teacher, Dr. McMullen observed that many students lack fluency in fundamental math skills. In an effort to help students of all ages and levels master basic math skills, he published a series of math workbooks on arithmetic, fractions, and algebra called the Improve Your Math Fluency Series. Dr. McMullen has also published a variety of science books, including introductions to basic astronomy and chemistry concepts in addition to physics workbooks.

Author, Chris McMullen, Ph.D.

SCIENCE

Dr. McMullen has published a variety of **science** books, including:

- Basic astronomy concepts
- Basic chemistry concepts
- Balancing chemical reactions
- Calculus-based physics textbook
- Calculus-based physics workbooks
- Trig-based physics workbooks
- Creative physics problems

www.monkeyphysicsblog.wordpress.com

MATH

This series of math workbooks is geared toward practicing essential math skills:
- Algebra and trigonometry
- Fractions, decimals, and percents
- Long division
- Multiplication and division
- Addition and subtraction

www.improveyourmathfluency.com

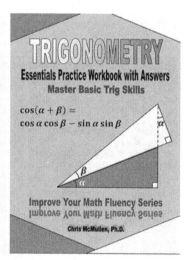

PUZZLES

The author of this book, Chris McMullen, enjoys solving puzzles. His favorite puzzle is Kakuro (kind of like a cross between crossword puzzles and Sudoku). He once taught a three-week summer course on puzzles. If you enjoy mathematical pattern puzzles, you might appreciate:

300+ Mathematical Pattern Puzzles

Number Pattern Recognition & Reasoning
- pattern recognition
- visual discrimination
- analytical skills
- logic and reasoning
- analogies
- mathematics

VErBAl ReAcTiONS

Chris McMullen has coauthored several word scramble books. This includes a cool idea called **VErBAl ReAcTiONS**. A VErBAl ReAcTiON expresses word scrambles so that they look like chemical reactions. Here is an example:

$$2\,C + U + 2\,S + Es \rightarrow S\,U\,C\,C\,Es\,S$$

The left side of the reaction indicates that the answer has 2 C's, 1 U, 2 S's, and 1 Es. Rearrange CCUSSEs to form SUCCEsS.

Each answer to a **VErBAl ReAcTiON** is not merely a word, it's a chemical word. A chemical word is made up not of letters, but of elements of the periodic table. In this case, SUCCEsS is made up of sulfur (S), uranium (U), carbon (C), and Einsteinium (Es).

Another example of a chemical word is GeNiUS. It's made up of germanium (Ge), nickel (Ni), uranium (U), and sulfur (S).

If you enjoy anagrams and like science or math, these puzzles are tailor-made for you.

BALANCING CHEMICAL REACTIONS

$$2\,C_2H_6 + 7\,O_2 \rightarrow 4\,CO_2 + 6\,H_2O$$

Balancing chemical reactions isn't just chemistry practice.

These are also **fun puzzles** for math and science lovers.

Balancing Chemical Equations Worksheets
Over 200 Reactions to Balance
Chemistry Essentials Practice Workbook with Answers
Chris McMullen, Ph.D.

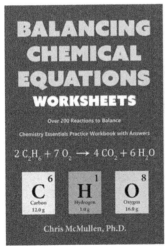
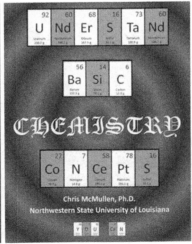

CURSIVE HANDWRITING

for... MATH LOVERS

Would you like to learn how to write in cursive?

Do you enjoy math?

This cool writing workbook lets you practice writing math terms with cursive handwriting. Unfortunately, you can't find many writing books oriented around math.

Cursive Handwriting for Math Lovers
by Julie Harper and Chris McMullen, Ph.D.

Made in the USA
Middletown, DE
13 September 2020